CEMENTING A PARTNERSHIP

THE STORY OF WAPCO 1960–90

CEMENTING A PARTNERSHIP

THE STORY OF WAPCO 1960–90

PETER PUGH AND J. F. ADE AJAYI

CAMBRIDGE BUSINESS PUBLISHING

First published in Great Britain, 1990, by Cambridge Business Publishing,
Cavendish House, Cambridge Road, Barton, Cambridge CB3 7AR

ISBN 1 871341 04 3

Designed by Geoff Green
Typeset by Goodfellow & Egan, Cambridge
Printed in Great Britain by St Edmundsbury Press

CONTENTS

AUTHORS' ACKNOWLEDGEMENTS

We interviewed many people and studied a great number of books and documents in our research for this book on WAPCO. We are very grateful to all those who helped us and hope they will forgive us if we just mention a few without whose assistance the book could not have been written and illustrated:

The architects, John and Jill Godwin, and the *Daily Times* of Nigeria lent many valuable photographs;

Dr Ransom and Dorothy Murray were charming hosts in Lagos;

Chief Ogunleye, Chief Jackson, Dr Murray, Oliver Nwaozuzu and Rodney Little read the drafts.

FOREWORD

BY LT. GEN ALANI AKINRINADE (RTD)
Hon. Minister of Industries 1986-1989

THE STORY of WAPCO told by two renowned historians who lived through the period runs like the industrial history of Nigeria since Independence. WAPCO has survived to narrate its experiences of a typical large scale industry passing through different periods of an agrarian economy – civil war, oil boom to gloom and doom, import licensing, economic stabilisation and finally Structural Adjustment. In the aggravated uncertainties, many industries collapsed but WAPCO expanded.

The company located itself on top of its major raw material, employed familiar and proven technology and stayed within easy reach of the largest consumer of its products. Only lack of linkage industries, especially capital goods, has prevented WAPCO from producing the cheapest cement in the world.

Our current Industrial Policy has set the stage for reasonable industrial growth and has created a sufficiently attractive investment climate for the next century. The abundance of raw materials, land, labour, hospitable climatic conditions is now crowned with a policy of economic liberalisation which was not available when WAPCO grew from strength to strength. It is now possible to invest foreign capital without unnecessary inhibition and employ the best management without undue bureaucratic intervention, conditions alien to our economic management in the past 30 years.

The company has been a visible and enviable landmark on the Nigerian industrial landscape when considering what is possible from the partnership of foreign capital, entrepreneurial flair, raw material and other resources. WAPCO must be a study for potential foreign investors in Nigeria and this book must be essential reading for the education of our people on our march to industrialisation.

I salute the founding fathers of WAPCO and congratulate them for their foresight, hardwork and tenacity of purpose. I feel very proud to be closely associated with the success of the company and commend the book to both public and private sectors.

INTRODUCTION

A A ABIDOGUN
Chairman of WAPCO

THE IDEA of writing the story of the West African Portland Cement Company Limited at 30 may appear unusual. The question may even be asked "why the story of WAPCO?" Thirty years in the history of a corporate organisation is probably of little significance in the sense that it is an organisation which will live in perpetuity. However, organisations, like human beings, are born. They grow into adolescence and reach maturity. Some will grow to old age and die. Yet others will continue to prosper until eternity.

It was decided to write the story of WAPCO because human memory is very short. For example, many of the founding fathers of the Company have died and it is only from records that we extract what we believe were the reasons for their establishing the Company. Others who are still living have only hazy ideas of the original intentions. For these reasons, we believe that 30 years were long enough to document the life journey of a Company of the size of WAPCO for the benefit of the general public. We also believe that given the track record of the Company, the story deserves to be told.

When it was decided to publish the story of the Company, time was not much in our favour. We have the privilege of having men of knowledge whom we trust would not find it too difficult to put things together, not only in terms of historical perspective but also in a manner that would be readable to the general public. The Company decided to invite Peter Pugh of Cambridge, a respected company historian, and our Prof J F Ade Ajayi, a world renowned historian and National Merit Award laureate, to write this story. The two between them have pooled a formidable wealth of knowledge and experience.

The story relates the genesis of the Company right from the

colonial era to the present day. It is indeed a picturesque narration of the vicissitudes of an organisation whose march towards fulfilment and achievement has been inextricably interwoven with the mutable fortunes of Nigeria. We are told how cement had been imported from the UK and other European countries from the latter part of the nineteenth century. The growth trend in the importation of cement showed clearly that it was a product which a country attaining independence should think seriously of manufacturing locally if it were to achieve the economic objectives of independence. Some of these, to my mind, are the achievements of self-reliance in vital economic activities, the promotion of employment opportunities and the acquisition of technical know-how.

It was not an easy task to locate suitable raw materials in sufficient quantity. In this connection, we should praise the pioneering efforts and indeed the courage and perseverance of the technical partners, then Associated Portland Cement Manufacturers Ltd of the UK, now Blue Circle Industries PLC. In spite of the lack of relevant data, in spite of the seemingly impossible terrain which they had to traverse and in spite of the hopelessness of the results initially achieved, they persisted and persevered until success came their way. The Company was therefore established at Ewekoro and it started production in 1960.

There were difficulties from the onset but the co-operation and the determination of the two partners, that is, the Western Region Production Development Board and the Associated Portland Cement Manufacturers Limited, surmounted the problems that arose both in terms of finance, the provision of facilities and the acquisition of land. There was no doubt that the establishment of the Company saved the country substantial sums of foreign exchange.

During its comparatively short history, the West African Portland Cement Company Limited has made tremendous contributions to cement availability in the country. Except during periods of depression and economic adversity, the demand for cement in this country has far outstripped its supply. Having regard, for example, to the slow down in growth and the depressed economy during the civil war of 1967–70 and the subsequent oil boom of the early seventies, Federal Government expenditure acted as a catalyst to the economy and cement produced within the economy became totally insufficient in relation to the level of demand. It was such a situation that led to the cement armada which occurred between 1974 and 1975 when the country ordered over 21 million tonnes. It was estimated that it would have required six years to discharge this quantity based on the capacity of the then existing port facilities.

Given the growth of the economy, the Company took action to increase its production capacity and hence, apart from expanding its works at Ewekoro, it also embarked on the construction of a new factory at Shagamu. The completion of that factory in 1978 took the combined installed capacity of the Company to about 1.6 tonnes per annum.

The economic fortunes of the country started to nose-dive in the early 80s. Government took physical and fiscal measures to cope with the economic situation. The regime of import licence was soon to start. The Company had difficulty in bringing in essential spare parts, gypsum and paper bags to cope with our operations. Output from our factories declined during this period. The local cement companies were only able to keep alive because the government responded to the call of local manufacturers by imposing a 20% import duty on imported cement. This tariff remained in force until the regime of the Structural Adjustment Programme started in 1986 and with the removal of the import licence scheme we were in effect operating a "cash and carry" system.

Apart from the physical development of the Company, there were developments in social welfare areas for workers and also in Nigerianisation. Many of the top positions in the Company are today manned by Nigerians and indeed WAPCO can proudly claim to be an example of an institution where the commanding heights are managed by Nigerians.

One area in which the Company has also excelled in line with the Nigerian Government's policy is the area of diversification. As far back as 1973, it acquired Cement Paints Nigeria, which became the Special Products Division of the Company, and later in 1986 a wholly owned subsidiary, Portland Paints and Products (Nigeria) Ltd. It manufactures decorative paints like Sandtex Trowel, Sandtex Matt and Sandtex textured and also cement based paints like snowcem and snowcrete. Another area of diversification is the establishment of the Portland Electrical Repairs Division, which repairs electric motors and alternators. Both the Portland Paints and Product (Nigeria) Limited and Portland Electrical Repairs Division are today making substantial contributions to the profits of the Company.

At 30, the factories of the Company now require refurbishment, redevelopment and modernisation. A substantial amount of money is required to do this. The Company will find it difficult, if not impossible, to face the challenges posed by these activities if we had to undertake them through self funding.

Our successful execution of these projects will require the active support of the Federal Government. Given the past records of the

Company, there is no doubt that all men of goodwill should rally round to ensure that both the present and the future of the Company is secured by giving us adequate support to ensure we continue to service the interest of the nation. We can proudly say that the last 30 years of our existence have indeed been spent in SERVICE TO THE NATION.

On a final note, I must salute all those who have served the Company in the past as Chairmen and Directors for their contributions which have carried the company so far. I regard myself as fortunate and privileged to be the present Chairman when the Company is marking this historic event of being 30. It is with great pleasure that I invite you to read this *Story of Wapco.*

1

PLANNING FOR DEVELOPMENT

The colonial background

The story of the West African Portland Cement Company (WAPCO) goes back to the twilight years of the colonial period, and the aspirations of Nigerian nationalists to achieve independence, and to end what they saw as the economic exploitation of the country. The slogan of the nationalist movement was to "seek ye first the political kingdom and all else shall be added unto you." High on the list of what they expected would follow from self-government was the creation of an Industrial Revolution, though they had no clear ideas

A typical expatriate residence in early colonial days in Force Road, Lagos

as to how this was to be achieved. During the colonial period local manufacture had been discouraged and the production of raw material for export encouraged. Thus Nigerians were encouraged to produce what they did not consume, such as cocoa, and to import much of what they needed to consume, such as flour and cement. Industrialisation was expected to reverse the process: agriculture would be improved to earn more foreign exchange; overseas investors would be encouraged to supplement this and set off the process of local manufacture of goods hitherto imported; this would produce employment and wealth, and promote the development of technological know how.

The colonial period of Nigerian history began in 1851 when the British Navy, ostensibly in pursuit of the crusade against the Atlantic Slave Trade, invaded and bombarded the island of Lagos. That opened the way for the establishment of a British Consulate, the allocation of land for missionary and commercial activities, and the immigration of westernised freed slaves (*Saros* and *Amaros*) from Sierra Leone, Brazil, Cuba and other places. This facilitated the continuing spread of missionary and British commercial interests into the Yoruba hinterland, the coastal city states of the Niger delta (Brass, Bonny, Kalabari), Calabar and the Cross River as well as the lower Niger regions.

In 1861 the British annexed Lagos, and the Consul became a Governor. In 1864 the Church Missionary Society consecrated Samuel Ajayi Crowther a bishop at the head of the Niger Mission staffed principally by Saros to open up the Niger basin to Christianity and Trade. Their success invited British, French and German firms after them in the 1870s. As the competition for staking out claims to African Territories hotted up in Europe, George Taubman Goldie united all the British companies into the United Africa Company in 1879 and sought a Royal Charter not only to protect his claims for commercial monopoly, but also to invest his company with powers of sole administration in the Niger Territories, i.e. Northern Nigeria and the Lower Niger Valley. He succeeded in 1886 and the Royal Niger Company emerged. By 1900 it was clear that the company was in no position to maintain its claims against the French and Germans, and at the same time enforce its rule on the various Nigerian peoples in the vast area. The British negotiated for a revocation of the charter and brought the Protectorate of Northern Nigeria under direct control, side by side with the Colony and the Protectorate of Lagos, and the Oil Rivers Protectorate. The latter two were reorganised in 1906 to form the Protectorate of Southern Nigeria, which in 1914 was amalgamated with Northern Nigeria.

In this piecemeal fashion Nigeria was gradually brought under British colonial rule and integrated into the world capitalist system in a position of dependency. The British administration existed principally to maintain law and order and promote the exploitation of Nigerian labour and natural resources, largely by British firms, for the advantage of the British. Its priority was to get Nigerian labour to produce raw material from agriculture and the mines for export and, through taxation and other means, promote the consumption of imported European manufactured goods. Internal trade and production for local consumption were neglected if not discouraged. The harbour and the railways running north and south were organised for the import/export trade. The major banks were foreign; they controlled Government deposits realised from taxes and customs duties paid by Nigerian peoples, but access to such funds was made difficult for Nigerian entrepreneurs. European firms and a few Levantine and Lebanese merchants with access to those funds advanced credits and loans to a few Nigerian distributors and produce buyers who best served their interests. Traditional crafts and manufacturing industries declined in the ruthless competition from European manufacturers protected by discriminatory regulations and practices; traditional mining for iron, tin, salt, galena, etc, declined; traditional textiles, leather goods, pottery, soaps, household utensils, etc, gave way for cheaper and more conveniently packaged European manufactured goods. Nevertheless, it was on the foundation of the colonial legacy, including the railway network, an agricultural economy and savings from the export of cash crops, that WAPCO was to be established at Ewekoro.

The Niger Company, even after it lost its charter, remained dominant in the import/export trade. As compensation for agreeing to give up the charter, it received a lump sum of £150,000 (worth about £2.25 million in 1990) and a promise of 50% of the royalties, rents and other charges to be levied on minerals in Northern Nigeria for 99 years. Between 1906 and 1944, this amounted to £2.25 million. In 1959, on the eve of independence, the UAC was persuaded to give up this claim in return for a compensation of a further £1m. With control of its own harbour facilities in Burutu, and its own shipping, it was a formidable opponent for any rivals. At the end of World War I, in 1920, its shares were taken over by Lever Brothers, a part of Unilever, the Anglo-Dutch conglomerate. By 1929, the Niger Company's main rival in Nigeria, the African and Eastern Trade Corporation, agreed to merge under the former name of the United Africa Company, with issued share capital of £15.7 million, wholly owning 37 companies and having controlling shares

The road to Obalende in the fifties

Ebute Ero 1954

in 20. It was reckoned as being responsible in 1946 for 43% of the import/export trade of Nigeria. An important item in the import trade was building materials, especially cement. The UAC had large wholesaling and retailing establishments; processing plants connected with the export trade, such as bulking plants for palm oil, clarified butter factories, and a sawmill in Sapele; 12,400 acres of rubber and palm oil plantations; a singlet factory in Lagos, and a number of cold storage plants. The Lever Brothers Soap Company in

Lagos was not a subsidiary of the UAC, but obtained its palm oil and kernels exclusively from the UAC.

By the end of World War II, the UAC was beginning to change its attitude and give more thought to the processing of raw material locally, though it remained cautious about the viability of such industrial proposals. Ahead of changes in government thinking, the UAC saw economic sense in moving away gradually from produce marketing and import distribution. In 1949, it added a plywood factory to its sawmill and by 1966 the African Timber and Plywood factory at Sapele had a capital of £3 million and employed a labour force of 3,200. Displaced from its market for beer in Ghana and the Congo by Swiss entrepreneurs brewing beer locally, the UAC moved to protect its market in Nigeria by establishing in 1949 the Nigerian Breweries Ltd. Then it took shares in Taylor Woodrow, a construction company with associated furniture-making companies. In this way, the UAC was pioneering in new import-substituting industrial ventures.

Regional Development Plan 1955-60

By 1954 the British had committed themselves to a transfer of political power to Nigerian politicians within the framework of a federation of three regions. Three major political parties emerged competing for power in the federation while each entrenched itself, especially among the majority ethnic communities of its region, and sought allies among the minority groups of other regions. The Action Group dominated the Western Region, especially in the Yoruba areas, as the National Council for Nigeria and the Cameroons (NCNC) did among the Igbo of the Eastern Region, and the Northern Peoples Congress (NPC) among the Muslim Hausa/Fulani/Kanuri of the Northern Region. There was great rivalry among the political parties as each became spokesmen and defenders of regional interests, particularly the interests of the majority groups. The British supervised the initial efforts of the Regional Governments, though inevitably not to the complete satisfaction of anybody.

The 1954 constitution took the devolution of power to the Regions to its widest extent. Each region was viewed as the base for development in education, health, agriculture and industry. The Nigerian Produce Marketing Board, which held the sterling balances from the profits of produce sold on the international market, was regionalised. The funds of the Regional Marketing Boards became the capital for the development of each Region. The formula for allocating revenue emphasised derivation above population and

The Queen and the Duke of Edinburgh with the Akarigbo, traditional ruler of Ijebu-Remo, during their visit to Nigeria in 1956

need. This favoured the Western Region because the cocoa, rubber and timber of the Western region earned more foreign exchange than the groundnuts, cotton and tin of the Northern Region and more than the palm produce of the Eastern Region. The days of crude oil were hidden in the future. The Civil Service and the judiciary were regionalised, but not the police or the army. The clause to allow for secession was rejected, but there was the promise that the Regions could achieve internal autonomy at their own pace. Thus the Western and Eastern Regions became self-governing internally in 1957, with the power to maintain Agents-General in London, while the Northern Region did not become self-governing until 1959, and the Federation in 1960.

The five year Development Programme of the Action Group was intended to take advantage of their increasing power over the apparatus of the state and, with the funds of the Cocoa Marketing

Apapa wharf from the Marina in 1954

Board, to lay a solid foundation for the development of the soon to be self-governing Western Region, especially in the areas of agriculture and industry. The Government was building up an effective Civil Service with manpower training and a scholarship programme, and preparing to launch a universal free primary education scheme and other welfare projects. The plan was to provide opportunities which had not been available in the colonial period for Nigerian entrepreneurs: access to bank loans, shipping and other infrastructure. More than that, by making available capital which individual Nigerian entrepreneurs could not have, the plan enabled Government to take the initiative in inviting foreign investors to bring their technology, management and skilled manpower in partnership with state capital to establish major industries as a base for the industrial development of Western Nigeria. Self-government provided the opportunity of new freedom to extend the search for foreign investors beyond Britain to include other countries of "the free world" with "close trade relations with Nigeria".

The Government established a Ministry of Development and two agencies, the Western Region Production Development Board (WRPDB, later known as the Western Nigeria Development Corporation, WNDC), and the Finance Corporation. How central the WRPDB was to the Government's development plans can be seen in the fact that it maintained a representative in London side by side with the representative of the Government, both with quasi ambassadorial status. The Economic Mission of March – April 1956,

Winds of Change – the Island Club in Lagos where Harold Macmillan made one of his "Winds of Change" speeches in the late fifties

sent to explore the initial projects of the Board, was headed by the Action Group leader and Premier of the Region, Chief Awolowo himself. Other members of the powerful delegation included the Minister of Development, the London representatives of both the Government and the Development Board, and the Chairman of the Board who was at the time still an expatriate officer.

The mission was away for five weeks and visited London, Washington, New York, Ottawa, Montreal and Toronto; then it split

Western Nigeria Commissioner, Chief Sowole, welcomes Chief Awolowo, J S Olawoyin, Rotimi Williams and others to London

Chief M S Sowole, the WNDC's representative in London

into two with a part visiting Dusseldorf, Bonn and Hamburg, Milan and Rome, while the other went to Tokyo, Osaka and Kobe in Japan. They found that overseas investors were interested but cautious. Industrialists knew that they had to respond to the changing political climate in order to protect their existing market in Nigeria. But it was far from clear that the political situation would remain stable, or that the cost of importing machinery and technology added to the local problems of infrastructure – labour and others – would not make goods manufactured locally too expensive to compete with imported goods. Such overseas industrialists needed to be persuaded with realistic feasibility studies that investments put into local production would be safe and in the end yield more than the profits from the export of manufactured goods.

Cement factor in development

To most communities in Nigeria, development in practical terms meant meeting the demand for pipe-borne water supply, electricity, roads, schools, markets, hospitals, post offices, etc. To a large extent, therefore, development implied building and building involved cement. The beauty of traditional towns and cities was in the design of the houses and the layout of the family compounds. For many Nigerians the fascination of European cities was in the grandeur of their individual buildings. Although there were some famous traditional multi-storey mosques, granaries and palaces using granite, clay and wooden rafters, the introduction in the twentieth century of the two-storey building with reinforced concrete first floor slabs was a landmark in many towns. It is not surprising that the 25-storey Cocoa House opened in 1965 was to become the symbol of the new age. Cement was the key to the building, not just of such monuments, but also of churches, mosques, multi-storey offices and modern private houses. The rate of the consumption of cement is sometimes listed not only as an index of the construction industry but also as an indicator of development.

It is not known when the first barrel or bag of cement was imported into Nigeria. Traditionally, houses were built of mud, or wood mortared with clay or a mixture of clay and lime. Wall surfaces were plastered, sometimes with treated animal dung, sometimes whitewashed with lime, sometimes polished with various preparations to resemble wood or black marble. Masons from Cuba and Brazil in the 1870s and 1880s revolutionized building techniques when they popularized the two-storey "Brazilian" houses and later built the Catholic and Anglican cathedrals and the Central Mosque in Lagos. They must have begun the import of cement. Europeans and the Saro merchants initially preferred to import their houses from Britain – prefabricated wooden structures, or iron frames lined with wood, sometimes adorned with imported brick chimneys, erected on iron girders. Initially, therefore, colonial officials imported cement more for building causeways, harbours, waterworks, railway stations and other public buildings than for private residences. The beauty and comfort of the "Brazilian" houses, however, soon persuaded them to adopt mud or concrete block houses plastered with cement.

Importation of cement through the Crown Agents, the Niger Company/UAC and other firms rose from about 6,000 tons in 1913 to about 40,000 tons in 1935 and continued to grow even in the 1930s, in spite of the world-wide economic depression. Sales dropped during World War II because of lack of shipping. In 1946, about 80,000 tons were imported. This figure was almost doubled to 154,000 tons by 1950, and quadrupled to 626,500 tons by 1960. The UAC was the largest importer, and most of the cement came from the Associated Portland Cement Manufacturers of Great Britain (APCM), subsequently renamed Blue Circle Industries. It is not surprising that cement was high on the priority list of the Development Board and that, in considering the feasibility of a cement project in Western Nigeria, they approached the UAC and the APCM.

The manufacture of cement requires heavy machinery and is therefore capital intensive. If limestone could be found in an adequate quantity and quality close to major roads, railways or waterways, it was considered worth exploring whether a foreign partner could be found to bring the necessary capital, technology and skilled manpower to produce and market cement locally. It would make an important contribution to development in Western Nigeria. Already the Federal Government had gone ahead with plans for a cement factory in the Eastern Region at Nkalagu which early surveys for limestone had indicated as the most suitable site on account of the quality of the limestone and proximity to coal mines. In 1950, The Nigerian Government approached APCM to establish a cement factory at Nkalagu. After nearly two years of surveying limestone resources in different parts of Nigeria, the APCM decided against the Nkalagu project because of the distance to the railway and remoteness from the Lagos-Ibadan area where most of their cement was sold at the time. The Government then approached a Danish firm manufacturing cement-making equipment and a British cement manufacturing associate to act as consulting engineers and management agent respectively in the Nkalagu venture, with the government finding the bulk of capital funds. The search for limestone deposits in Western Nigeria was as yet inconclusive. Initial reports suggested that what was abundant was kaolin, not limestone, but a geologist financed jointly by the APCM and the Development Board was investigating the matter further. The Economic Mission in 1956 discussed with the APCM in London the possibility of establishing a cement factory to import clinker and grind it or, if limestone deposits proved adequate and suitable, to manufacture clinker and grind it locally. In such a venture, while the Development Board would invest some capital and provide local facilities, the APCM as technical partners were offered majority shares so that they could demonstrate a long term confidence in the project through such investments.

The politics of self-government

As discussions for the establishment of WAPCO went on, the march to independence was producing significant changes in the political environment of Nigeria. The main political parties having consolidated their positions, each in its own region, began to shift attention to the structure and the control of the federal government. There was some retreat from the extreme regionalisation of the 1954 constitution. The capital city of Lagos which had been regarded as an integral part of the Western Region, on the outskirts of which the Action Group Government was building an Industrial Estate, was

Princess Alexandra with the Governor General, Dr Nnamdi Azikiwe, and the Premier of Eastern Nigeria, Dr Michael Okpara

detached from the Region and administered directly by the Federal Government. A clause was inserted in the revised constitution to empower the Federal Government to declare an emergency, suspend the government of a Region and administer it directly if the need arose. Although residual powers remained with the Regions, it was becoming clear that the powers allocated to the Federal government both on the exclusive and concurrent lists were substantial and likely to grow more rapidly than the power of the Regions. In particular, the powers of the Federal Government over banking, customs and excise, company tax, and overseas borrowing were such that the Regional Government could no longer ignore the Federal Government in its development schemes. It was the Federal Government that was preparing a package of incentives for foreign investors, including five years tax relief for companies granted pioneer status on condition of some Nigerianisation of their personnel; import duty relief on machinery, spares and raw

The Prime Minister of the Federation of Nigeria, Alhaji Sir Abubakar Tafawa Balewa, with Prime Minister Harold Macmillan, just before the Commonwealth Prime Ministers' Conference in 1962

materials; initial capital allowances and depreciation allowances for equipment; and protection of local industries through tariffs, quotas and licensing. The First National Development Plan did not come until 1962-68.

The Prime Minister of the Federation in the transitional period was Alhaji Tafawa Balewa. He was deputy leader of the NPC, the leader being Alhaji Ahmadu Bello, the Premier of the Northern Region, and Sardauna of Sokoto. Balewa owed his position to the allocation of majority seats in the Central legislature to the Northern Region. In preparing for independence, the leaders of both the Action Group and the NCNC resigned their positions as Premiers of their Regions, leaving their deputies in charge, while they tried to win power in the Federal legislature. After the elections in 1959, the NPC had the largest number of seats, but not enough to form the Government. Both the NCNC and the Action Group sought alliance with the NPC so as to share in the government. The NPC preferred the NCNC. Thus Balewa was confirmed as Prime Minister and Dr Nnamdi Azikiwe, the NCNC leader, became the Governor-General. Chief Obafemi Awolowo became the leader of the Opposition, and the Action Group faced a bleak future. While its hold on Western Nigeria seemed secure enough to reassure investors, its place at the national level, within the context of intense competition among the rival political parties, was a factor of potential instability.

2

THE DECISION TO INVEST

Would cement manufacture be justified?

By the mid fifties Nigeria was providing a significant export market for Associated Portland Cement Manufacturers (APCM). Indeed, in a report to the board in February 1955 the APCM export director, Michael Chapman, spoke of the company's

> "inability to give the market all it wants... Consumption of cement in Nigeria is still increasing. It is likely that most of the provisions in the International Bank's Report [on Development in Nigeria], amounting to an expenditure of £400 million over the next five years, will be carried through which will build up consumption to 400/500,000 tons a year."

Of all import-replacing industries none possesses so great a comparative advantage as the production of cement. The saving on sea transport is enormous particularly with reference to its low value and in the fifties the tariff on the import of cement was 9-16% of the cif value. Furthermore there was the clearance of bagged cement through the docks and the inland freight. Cement was an obvious product for the Nigerian Government to consider for local manufacture. As noted above the Nigerian Government approached APCM to establish a local plant, but when APCM declined the offer turned to the Danish company, FL Smidth and its British associate, Tunnel Cement, who acted as consulting engineer and managing agent respectively in a plant erected at Nkalagu in the Eastern Region. In this venture the Nigerian Government provided 68.6% of the equity capital of £1.75 million and 100% of the loan capital of £500,000. Tunnel and Smidth only invested £93,750 each.

In the fifties cement was shipped from England by sea, in bags – this shipment is being loaded at Bevans Works on the Thames in Kent

The demand circumstances, allied with the general feeling that there was some obligation for Britain to help create manufacturing companies in Africa, led APCM into talks with the Western Regional Production Development Board of Nigeria (WRPDB) to carry out surveys to determine whether the right raw materials existed in Western Nigeria to make a cement plant viable. APCM must also have felt reluctant to let another important market fall to a competitor. However, Michael Chapman was not optimistic,

"The possibility of finding suitable raw material in the right areas is not at all promising according to present

> geological evidence... There is said to be a good deposit at Ukpilla in the Kukuruku Division of Benin Province, but this is 'in the blue' and would appear at first glance to be impossible. It is nearly 300 miles by road from Lagos (which would probably cost 100/– a ton to transport) and about the same distance from the nearest point of the Railway and the main consumption areas of the Western Region."

In spite of this pessimism the boards of APCM and WRPDB signed an agreement on 27 May 1955 whereby a survey of limestone deposits would be carried out in Western Nigeria with a view to establishing a cement factory. Both parties would share the cost and the information. WRPDB would retain the required Prospecting Rights and Exclusive Prospecting Licence, but would nominate APCM as their agent. The agreement concluded,

> "In the event of either party deciding to complete the investigation or to erect a cement factory it will invite the other party to participate. ... In the event of one party continuing alone (the other party having declined the invitation to participate) it will acquire full ownership of the technical reports after repaying the other party for its share of the cost of the reports."

APCM undertook to engage the survey team and assume full responsibility for its passages, pay, tentage, equipment, etc. The expected cost of the survey was £20,000 (which sounds very small, but in today's money would be about £300,000).

There must have been some caution on the part of WRPDB in making an agreement with APCM as the British company had declined to proceed with the construction of a plant at Nkalagu in the Eastern Region in the early fifties after signing a Letter of Intent with the Federal Government. Indeed John, now Sir John, Milne remembers that he and Tom Chesterfield were given a frosty reception by Ronnie Berrif, a colonial official, formerly in the Eastern Region, when they visited him in Kaduna in their capacity as overseas representatives of APCM considering the possibility of a plant in Nigeria.

The search for limestone

Following the agreement two employees from the APCM research department, Harry Pollitt (later general manager of Research

Nigeria

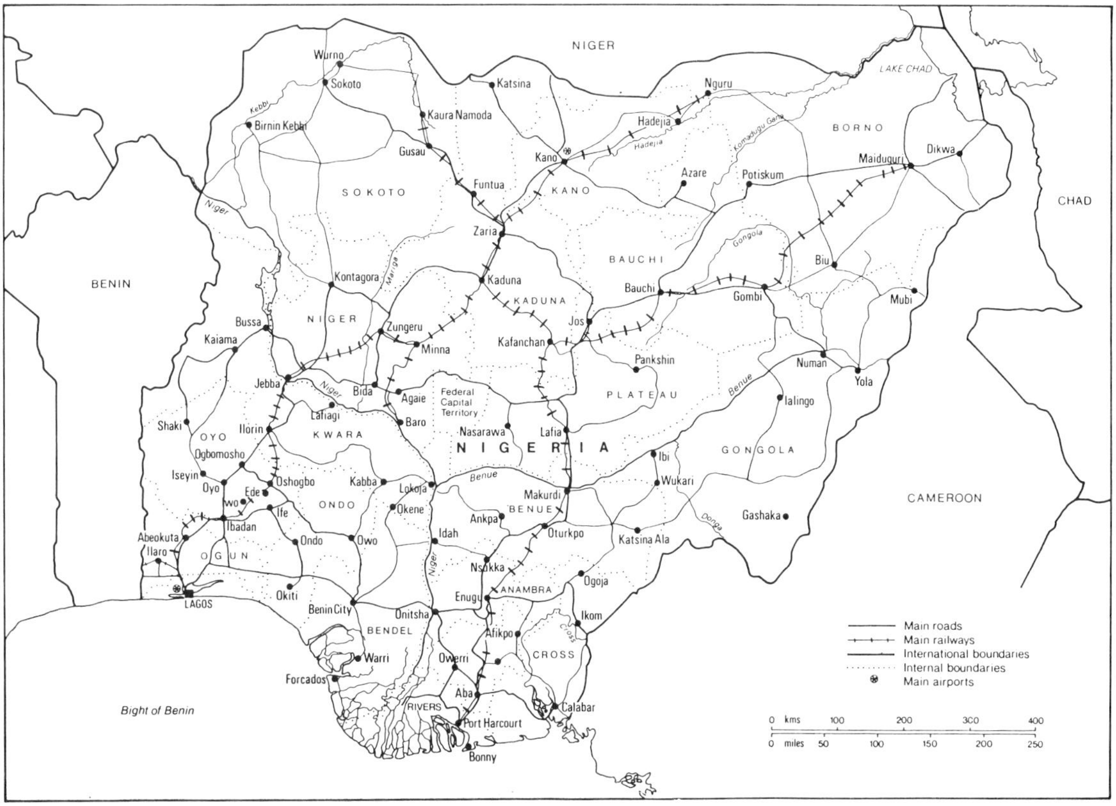

A map of Nigeria with the main roads and railways

Department), and Barry Dales met two members of the Directorate of the Colonial Geological Surveys, Dr Dixey and Dr Varley, in June 1955 and discussed the results of previous surveys. According to Dixey, the latest information suggested that the two outstanding sources of limestone in Nigeria were at Nkalagu in the Eastern Region and Jakura in the Northern Region. Dixey's view of the Western Region was that,

> "Investigation of the southern tertiary sediments has so far revealed only a few limestone boulders among the sands and clays, which were of academic interest only and in his view *it was extremely unlikely that a limestone deposit of economic size would be discovered there at this late date.*"

In July Dr Jacobson, director of the Nigerian Geological Survey, was invited to the APCM research department and met APCM's director of research, Edward Burke, as well as Pollitt, Dales and Cliff

Robinson. He confirmed the views of Dr Dixey – "Dr Jacobson considered it most unlikely that any limestone deposit suitable for a cement works would be found in these young sediments." There was certainly limestone around Ukpilla in the north of the Western Region, but it was very badly placed to supply the area of greatest demand around Lagos and Ibadan. Jacobson did admit, however, that a systematic detailed geological mapping programme had not been carried out in the Western Region. Nevertheless,

> "During the recent search by the Survey in Nigeria for limestone deposits suitable for cement manufacture attention had been given to the more populous areas of the Western Provinces and in addition all new excavations such as road cuttings etc had been examined *but no limestone had been discovered as a result."*

It was suggested that aerial photography might be useful in the absence of comprehensive maps and that the oil company Shell D'Arcy had been very active and successful in this field in their search for oil. As well as study of these photographs Jacobson suggested that a visit to Kaduna to study the results of the Colonial Geological surveys would be helpful, but he was definitely not hopeful of APCM finding economically viable quantities of limestone in the Western Region.

Undaunted, or perhaps daunted but undeterred, Pollitt and Robinson flew to Nigeria and in August 1955 wrote to Burke from Ibadan reporting progress after their visit to Kaduna. The one good thing they had learned was that the southern part of Western Nigeria had not been properly surveyed and that therefore the negative conclusions they had been given were not necessarily justified. They learned that Shell D'Arcy had spent the staggering sum of £8 million (probably £120 million in today's terms) producing a geological map of southern Nigeria, but even so had only covered about half of the Western Region.

By 28 August Pollitt had visited parts of the Western Region and was beginning to sound mildly optimistic,

> "The length of the tour here has been extended because we have had a clue from the Agricultural Research people at Ibadan suggesting a deposit near the Dahomey border to the West of Abeokuta (on the main road north of Lagos) and in view of the favourable position we feel we must investigate it. The deposit, if it exists, is unknown to the Geological Survey here so that it would be doubly satisfying to find it, again if it exists."

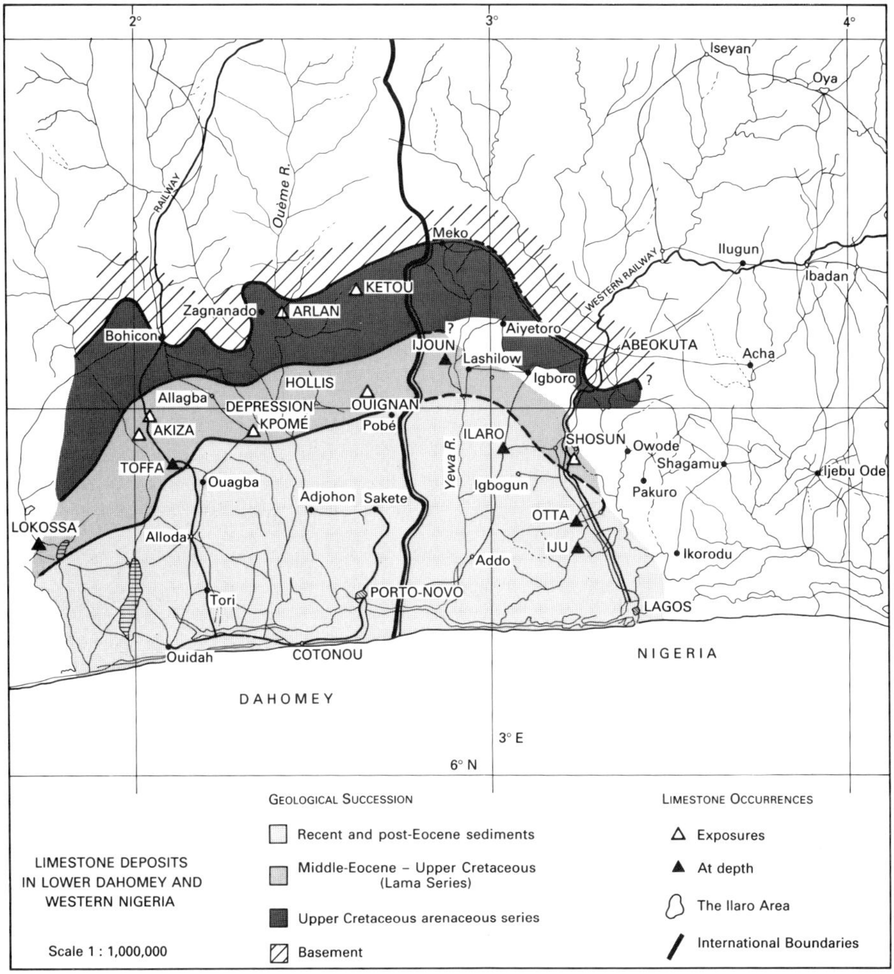

The limestone deposits in the former Lower Dahomey and Western Nigeria

By 5 September considerable progress had been made as Pollitt had travelled to the border of Western Nigeria and Dahomey (at the time a French colony, now the independent state of Benin). Pollitt reported,

> "Yesterday we went to the French customs port at Idiroko to ask whether anyone knew anything about the newly reported limestone in Dahomey. To our amazement the Customs Officer invited us to go and see the Chef de Division at Ketou, who he thought might know something about it. So we spent the next few hours in Dahomey and located surface deposits of limestone debris similar to that at Ijoun. The Chef de Division also marked a cross on our map to show the site of the newly found deposit and gave us 20 litres of petrol to make sure we didn't get stranded. All this without passports, international driving licences and so on! Vive

> l'Entente Cordiale! We couldn't in the time get as far south as the deposit in question but if it occurs at all in Western Nigeria it might be found in the region of Ilaro (North West of Lagos)."

What Pollitt did not say in his report was that his fluency in French was almost certainly a great help. In his concluding report he promoted the Ilaro area as the most promising. Ukpilla had satisfactory deposits of limestone but the quality was dubious and its situation relative to the market place totally unsatisfactory. There was also limestone in the Owan River Siluko area near Benin City but it was likely to be contaminated with sand and therefore in need of special processing.

> "The Ilaro area is ideally situated but the quantity and quality of limestone are unknown. There is the possibility of an extension of a reported French discovery of limestone in Dahomey."

He recommended that this last area be investigated further and that a series of exploratory holes be drilled. Pollitt's report was studied by the boards of APCM and WRPDB and on 23 December RA Crofts, the chairman of WRPDB, wrote to Arthur Poole, the managing director of finance and overseas development of APCM. He commended the excellent manner in which Pollitt's report had been prepared and accepted APCM's suggestion that a more detailed reconnaissance of the area from Ilaro to the Dahomey border be carried out.

On 1 February 1956 Pollitt and Michael Spurr of APCM met Chief MS Sowole and L Hood of WRPDB and agreed the formula and details of the further survey. By 14 February Michael Spurr recommended the following plan of action,

1 Visit the outcrops of limestone at Arlav, Ketou and Ovignan in Lower Dahomey. This would provide some knowledge of the nature of the geological succession that they would hope to penetrate by drilling in Nigeria.

2 Sight and drill initially a single 400-500 foot borehole near Ijoun, the site to be determined by information gained in Dahomey.

3 Carry out a ground geological survey, supported by a series of shallow boreholes, in the area of the

headwater streams of the River Yewa south and east of Meko.

4 Further deep drilling near Ijoun.

By 3 March Spurr was in Ibadan and was making his plans to visit Dahomey and then to return to Jos in Nigeria to obtain a prospecting licence. And off went Spurr to Dahomey and wrote back to Burke on 4 March, "So far I am keeping very fit and am only too anxious to begin my search" though he was not finding living conditions in Dahomey much fun.

> "The price of everything is outrageous! Nearly all the food is imported, even meat and potatoes. A mediocre lunch costs 21/– [about £16 in today's money], a bottle of beer over 5/– [£3.75] a glass of Perrier 2/– [£1.50] and a pound of butter 10/– [£7.50]. To live at the only hotel in Cotonou costs over £4 [£60] a day, and it's no Ritz either! However, the snails I had for dinner last night weren't too bad!"

By the end of March Spurr was back in Nigeria with useful information from Dahomey and on 25 March opened his report to Burke with the simple statement,

> "Limestone has been found in this area."

He went on to say that the rock had a very shallow dip southward, appeared to be 30 to 40 feet thick and traversed the area in an east, south-east direction from the Dahomey border to beyond the Western Railway of Nigeria. The deposit was exposed on either side of the tarmac road from Ilaro to Abeokuta nine miles north of Ilaro station and also at a point on the main Western Railway two miles south of Itori station. Spurr was already able to say that the limestone quantity was sufficient to support a single unit cement works. He also expressed his surprise that the limestone had not been discovered earlier. He admitted his luck in Dahomey,

> "You may be wondering how I knew more details about the Dahomey succession than I did in London. Well, on Monday in Cotonou, I had a terrific stroke of luck. When in the Mines Department the post arrived. In it was a report of the borehole sunk near Pobé. M. Sifre put it on one side but I persuaded him to let me have it for an hour. It was supposed to be confidential. I made notes

> and that was that! I made haste to Ilaro and began my search."

Spurr was already looking at Ewekoro as a possible site for the cement works,

> "As suggested on the map of the area, which you should now have received both the road and railway cross the Ewekoro area by means of embankments."

However, in a letter to Burke two days later Spurr wanted to look at other areas,

> "As soon as the drill and the drillers arrive I shall deploy them in the Ilaro area. I consider that any further work in the Ewekoro area should be postponed for the time being."

In a letter to Spurr of 24 March Burke was full of praise using phrases such as "incredible but very acceptable ... very pleased with yourself and rightly so." At the same time Burke warned him to be cautious until the samples had been fully analysed,

> "You must on no account say anything which might lead the local people to assume that we have succeeded in our task as yet and raise hopes that might later prove unfounded. In particular you should avoid discussing questions of compensation which could well lead to trouble until that trouble is known to be inevitable."

Both Burke and Geoffrey Davis, APCM's works managing director, were anxious that Spurr should not rush headlong and sacrifice the advantages gained by his rapid and successful discoveries. They also insisted that their partner, WRPDB, was kept fully informed. Spurr got the message and wrote to the WRPDB on 9 April outlining his progress and telling them of the finding of limestone deposits north of Ilaro and near Ewekoro on the Western Railway. Drilling was continuing to determine the choice of the best area for more detailed investigation. Disturbance to the local population would be kept to the minimum possible and if it was necessary to drill on private property it was hoped that the board could inform the local farmers of the nature of the investigation and that compensation for any damage would be paid by the company. The WRPDB responded well and the acting chairman, AO Rewane, wrote to his Highness Sir Ladapo Ademola II, the Alake of Abeokuta, outlining the plans of the board and APCM to build a cement works and asking for his co-operation.

By June 1956 Spurr was ready to return to London. He felt his task was complete and that the monotonous sinking of boreholes could be left to others,

> "It is very tedious waiting for boreholes to be put down!"

Norman Brindle (far right) with T Hick and A Ryan. He was one of APCM's geologists sent to follow up Michael Spurr's initial survey

Brindle explaining to Oba Akran, the Western Nigeria Minister for Development, how surveys are carried out with the help of aerial photographs

Aspects of geological investigations

One of the geologists who replaced Spurr was Norman Brindle. Spurr had found limestone at Ilaro and Ewekoro but was there enough to justify the major investment of a works? Norman Brindle was one of the team that surveyed for the next 18 months to ensure that there was sufficient recoverable limestone. Brindle used aerial photographs instead of maps as the only maps available were relatively small scale whereas the photographs were 1:30,000 and made it quite easy to locate drilling sites. There had been an RAF station at Jos during the Second World War and afterwards the RAF had flown and photographed the whole of Nigeria. Brindle found these photographs extremely helpful as surveying by traditional methods would have been time-consuming and expensive.

Preliminary decisions

By April 1957 the decision had been made to build the works at Ewekoro which was seen to offer the optimum site in terms of quantity and quality of limestone and location to the large market of Ibadan and Lagos with good road and rail links. There were some drawbacks to the Ewekoro site. For example, the limestone was beneath the water table. Edward Burke was undeterred by this, pointing out that the same problem had been overcome at the APCM works in Malaya. It was also not going to be easy moving the extensive overburden in wet weather. Two engineers from APCM's Works Department, Philip Willmott and OW Jones (always known as OW), were sent out to Nigeria to,

> "obtain further information as to the relative merits and cost of coal versus oil as fuel for the works, to discuss arrangements and terms for the supply of electricity and to make arrangements for the acquisition of the necessary lands."

They were also asked to obtain information on the proposed site and on appropriate construction methods and standards and to make contact with authorities such as the railways, ports authority and customs.

After discussion between Willmott and Jones of APCM, the Hon. Alhaji DS Adegbenro, the Minister for Lands and Labour Western Region, Mr Rewane and Mr Whyte of WRPDB, it was decided that, rather than have prolonged negotiations with local landowners, whose title was often very obscure, powers recently granted in the Assembly could be used to acquire in one operation all the land that might be needed then or in the future. This would obviate any feelings of unfairness that might arise from a multiplicity of dealings. The Government would then lease the land or vest it absolutely in the cement company. WRPDB would compensate the landowners and would be subsequently reimbursed by the cement company.

The Hon. Alhaji DS Adegbenro enquired about the availability and training of Nigerians to work in the company. He was assured by the APCM representatives that wherever possible local people would be employed and that the company intended to train fully key operatives and staff. It seemed clear that general operatives could be recruited from local villages, but the skilled grades would need to be filled from Lagos and Ibadan and that these operatives would need to be provided with housing. Clerical labour would also have to be brought in from the large towns and housed in a style commensurate with their status. The standard working week was 44 hours, although there were moves afoot to have it reduced to 40 hours. There would be no objection to shift working and a four shift as opposed to a three shift system was recommended. The cement works at Nkalagu was already operating a four shift system.

Discussions were held with the Nigerian Railway Corporation, the Nigerian Coal Corporation, the Mobil Oil Department of United Africa Company and the Texan Petroleum Company, representing African Petroleum Terminals, and also with the Electricity Corporation. Following these it appeared that the most economic fuel would be oil.

The Nigerian Railway Corporation was the sole selling agent for the Coal Corporation and proved to be unwilling to make any price concessions for bulk buying. The Coal Corporation gave the impression "that they would be pleased to supply us with coal but would not be the least bit worried if we decide to use oil".

The Electricity Corporation, a large consumer of coal, complained of poor and variable quality and of bad weights and poor loading of wagons resulting in loss by spillage. As regards the supply of electrical power to the works they foresaw no problems.

> "With regard to private generation, Mr King, the general manger, said he would not raise any objection if we wished to generate, but he thought we could not do so more cheaply than ECN could supply. He advised us to go very carefully into this as he knew of another firm who had tried it and now wished to take a supply from ECN instead."

Apart from talking to the Nigerian Railway Corporation about the supply of coal the APCM engineers also discussed with the corporation the supply of materials to and from the works. The main Lagos/Ibadan line ran just east of the works site and the Railway expressed its willingness to shunt full and empty wagons in and out of the works and to lay out sidings. After examining the drawing of the kiln and mills, the Railway said that the transport of kiln sections was not possible over their lines, but there would be no problem with the smaller pieces and they foresaw no difficulty with gypsum, cement, spare parts or stores.

Nor did the Nigerian Ports Authority pose any problem although the manager, CF Wallace, suggested that it might be quicker and cheaper to break the kiln down into sections. It was also suggested that for ease of identification and speed of clearance it might be advisable to have an APCM man on hand who would recognise the parts. Routine shipments of gypsum could be loaded direct into rail trucks and not dumped on the ground.

The engineers also went to see the Controller of Customs to enquire whether there would be any problems over the import of equipment of East German origin. (The kiln was bought from Polysius, a German cement company. Polysius was run by a fascinating character called Schmidt, who had left Germany in 1935 because he could not stand the Nazi treatment of the Jews. Changing his name to D'Arcy Smith he served in the RAF during the war where he was extremely valuable in being able to advise on appropriate industrial targets. After the war he returned to West Germany and it was from him that APCM bought the original kiln for Ewekoro works. The plant had originally been destined for a projected cement works in Newfoundland, Canada.) Willmott and Jones also wanted to know if there was the possibility of preferential tariffs in respect of raw materials imported for the construction of the works. An import licence would be required for imports from Germany. There would be no duty on machinery although there was a 20% *ad valorem* duty on building materials. However, there was the possibility of a subsidy.

Willmott and Jones then studied the road between Apapa docks at Lagos and Ewekoro and recommended that all heavy loads be transported during the dry season, between 1 December and 10 April.

On the site of the works itself, they wanted to wait for the report of Richard Costain Ltd, the chosen contractor, but recommended that the design and construction of the main works buildings and the machinery foundations, ie the kiln house, the raw materials and clinker store, the mill house, the slurry basins and elevated slurry tanks, the packing plant and the cement silos, should be the same as in Britain. Sixteen houses for Grade 1 junior staff, 20 houses for Grade 2 junior staff and 93 houses for skilled and semi-skilled workers were to be provided as well as a small dispensary and a community building for games and cinema shows. It would also be necessary to provide football pitches and tennis courts.

The area for senior staff housing at Ishofin was three miles from the works on a hill beside the Lagos road. The house proposed for the manager was two-storeyed with three bedrooms and servants' quarters. (In the event it was not built as a two-storey house.) For the eight senior staff, bungalows along the lines of those provided for UAC managers were suggested. A temporary block of 18 single quarters was recommended for those brought out to train the Nigerians in the works. A club house would be provided with lounge, bar and room big enough for dancing and cinema shows.

3

THE BUILDING OF EWEKORO

Heads of Agreement

By early 1957 the three parties that had discussed the joint venture of a cement manufacturing plant in Western Nigeria were ready to take a firm step to bring the venture to fruition. On 17 May 1957, APCM, WRPDB and UAC signed a Heads of Agreement.

They agreed, subject to the following conditions,

> Satisfaction with the outcome of investigations in progress
>
> Agreement by all parties on the method of implementation whereby APCM would carry out the necessary design and engineering services in connection

Prime Minister Tafawa Balewa, Commissioner Sowole and the last British Governor General of Nigeria, Sir James Robertson in the UK in the early sixties with the construction of the plant and UAC would provide commercial and administrative services

The obtaining of necessary permission from the Capital Issues Committee to the establishment of a cement manufacturing plant in Western Nigeria.

These were the proposals;

Based on an output of 100,000 tons and assuming a capital cost of the order of £2.5 million, the three parties agreed to participate as follows:

APCM	51%
WRPDB	39%
UAC	10%

If investigations show that a plant with a greater capacity would be justified the three participants agree in principle to consider making further contributions to the capital.

A new company to be formed with registered office in Nigeria. The name of the new company to be "West African Portland Cement Company Limited".

The Board to consist of six members nominated as follows:

APCM 3, WRPDB 2, UAC 1.

The Chairman to be a nominee of APCM with casting vote. The structure of the Board to be reconsidered in the event of any alteration in the shareholding of any party.

The APCM to provide all the necessary design and engineering services in connection with the construction of the plant and to exercise full technical supervision over the production processes on terms to be agreed.

The UAC to be responsible for all work in connection with the formation of the new company and to provide commercial and administrative services to the company when formed on terms to be agreed.

Preliminary out-of-pocket expenses incurred by any of the parties, including cost of company formation, to be

> recovered from the new company. Any preliminary expenses not covered by the Agreements between the APCM and the WRPDB to be agreed by the three parties and to be recovered from the new company. In the event of the company not being formed any preliminary expenses outside the existing agreement between the APCM and the WRPDB to be borne by the party concerned.
>
> In view of the large capital expenditure involved the WRPDB to explore the possibility of obtaining the exclusive right to mine or quarry materials for cement manufacture, to grind clinker and to manufacture cement in the Western Region.
>
> The WRPDB to endeavour to obtain an assurance from the Federal Government that they will not participate in or in any way support a clinker grinding project in the Federal territory.
>
> The WRPDB to approach Federal and Regional Governments in respect of fiscal policy.
>
> The new company to undertake training of Africans in new skills and techniques.

The agreement was signed by John, later Sir John, Reiss, the chairman, and Geoffrey Davis, a managing director of APCM, by RA Crofts the chairman and AO Rewane and Chief MS Sowole on behalf of WRPDB and by AN Smith and ACC Baxter on behalf of UAC.

WRPDB had persuaded APCM to invest in Nigeria and were happy for the company to have a majority shareholding. The attitude of WRPDB was that there was an element of risk and that therefore they should subscribe for the shares rather than members of the Nigerian public. The plan was that once the element of risk was reduced the public could take over their shareholding.

Construction of Ewekoro works

Construction began in the middle of 1958 when the first contract, which was for staff housing, was awarded to TA Oni and Sons. This was quickly followed by contracts for piling, civil engineering works and additional housing awarded to Frankipile, Gammon (Nigeria), Cappa & d'Alberto and Costain (West Africa). The Electricity Corporation of Nigeria provided supply by the special construction of a 40 mile overhead transmission line from Ijora. The Nigerian Railway Corporation laid down works sidings adjoining the main line.

The erection of the plant was carried out by staff of APCM. Harvey Robinson, who had managed a cement plant in Burma and had been forced to destroy it to prevent it falling into the hands of the Japanese during the Second World War, was appointed construction manager. It was one of APCM's first constructions of a cement plant overseas and apart from one or two mistakes – the holding down bolts on the Davis pre-heater were not grouted in properly causing a two week delay in commissioning – great credit must go to the team which built up an efficient works in difficult terrain in a country without an industrial background. The APCM team was supported by supervising engineers from Vickers Armstrong, GEC, Brown and Tawse, and Ruston and Bucyros. Whessoe designed and erected the oil storage tanks. APCM staff also carried out the electrical installation with assistance on the erection of the electrical switchgear from Reyrolle. The installation at the housing estate was carried out by the Nigerian Electrical Contracting Company and Utilities (West Africa). Palm Line Agencies handled the import of the machinery and it was transported to Ewekoro by Bank's Transport.

By the time the works was completed it was the largest single manufacturing establishment in Nigeria.

The commissioning of Ewekoro

The first chairman and managing director of WAPCO was Tom Chesterfield. He had started in the APCM mailing room before the Second World War on the princely wage of 15 shillings (75p) a week. During the War he served in the Royal Tank Regiment in the Middle East, North Africa, Italy and North West Europe. After the War he rejoined APCM and after appointments in the sales department joined the export department in 1952 and moved to Accra in the Gold Coast as the company's sales manager in West Africa.

On 3 December 1960, five-and-a-half years after APCM and WRPDB had made their agreement to search for limestone, the works at Ewekoro was commissioned by the Governor of the Western Region, Sir Adesoji Aderemi. John Reiss had flown out from the UK for the occasion. There was a moment of horror when Tom Chesterfield and John Reiss were driving from Ibadan, where they had stayed the previous night to meet various dignitaries, and were held up by a tree that had fallen across the road in the heavy rain. This setback was overcome, but they were faced with another when

John Reiss, chairman of APCM, welcoming Chief Awolowo, the Premier of the Western Region, to the foundation ceremony for Ewekoro Works. With him are Mrs Awolowo and D S Adegbenro, Minister of Local Government

The Chief signs the visitors' book at the foundation ceremony

Ewekoro –
– limestone tipping to crusher
– "A" type bungalow, senior housing estate
– "B" type bungalow, senior housing estate

Ewekoro –
– cake feeders, from Clarke-Chapman conveyor
– oil storage tanks
– rail sidings

Ewekoro –
– laboratory and dispensary buildings
– raw materials store
– limestone
– hard tennis courts, senior housing estate

Ewekoro –
– limestone conveyor from quarry
– Davis pre-heater
– raw mill

Ewekoro –
– Fluxo packer in operation
– slurry storage plant

Ewekoro –
– Blue Elephant Club
– main offices
– quarry

Ewekoro –
– Works community centre and sports ground, junior housing estate
– filter presses
– Clarke-Chapman conveyor looking up to pre-heater

Ewekoro –
– fitting shop
– main stores
– paper bag storage
and packing plant

Ewekoro –
– swimming pool, senior housing estate
– Tom Chesterfield the first chairman and managing director of WAPCO, (right) with Arthur Poole, managing director APCM Overseas

the kiln would not light for the commissioning. This further potential disaster was avoided by the simple expedient of setting fire to some oily rags and blowing the smoke up the chimney. The works had cost £4 million (about £60 million in 1990 terms) and had transformed the little village of Ewekoro. As we have seen the original geological surveys had suggested the limestone deposits would not support a cement factory, but the APCM/WRPDB investigations had discovered resources that would support production for a hundred years. The construction of the factory had occupied 80 expatriate Britons and 1,800 Nigerians for two and a half years. The *Nigerian Daily Express* pointed out that,

> "The main features of the Ewekoro factory are a 500 horse power hammer crusher with a crushing capacity of 200 tons of limestone an hour; an oil-fired kiln 200 feet long, twelve feet six inches in diameter, weighing 600 tons; two 1,100 horse power cement grinding mills and two silos, 90 feet high and 35 feet in diameter with a storage capacity of 2,500 tons each. Its chimney will rise over the surrounding area to a height of 150 feet."

The company hoped to produce 200,000 tons a year and would market the cement under the brand name 'Elephant'. Clearly the company had chosen the African symbol of strength. The company emphasised that the Elephant cement produced at Ewekoro would be of the same quality as the Burham cement manufactured by APCM in England and supplied to Nigeria since the 1880s.

The *Nigerian Daily Telegraph* saw great significance in the opening of the Ewekoro plant,

> "We are happy that Ewekoro factory has been opened because, not only will it supply much needed cement, but it will also give employment to the jobless and thus help minimise unemployment in Nigeria. As Chief Akintola said on Saturday and as we have always said, political battle is over and the next battle is economic. Every industrial establishment opened is one theatre of war won. This is grand battle because it is the battle of the stomach."

What process was used at Ewekoro?

The raw material that had determined the choice of Ewekoro as the site for the works was a mixture of limestone and soft shale. The shale lay above the limestone and two thirds of it was removed exposing the limestone for blasting. Two five cubic yard RB110 dragline cranes were used, one for removing the shale and the other for mining the limestone. Both materials were removed by 9 cubic yard Aveling Barford dumpers. The red shale was in a 7 foot layer beneath the top soil and underneath it was a 23 foot bed of grey shale which was discarded. The limestone was in a bed 35 feet deep, below the grey shale, and at its base was 66 feet below ground level.

The limestone's preliminary crushing was done in a Dixie swing hammer-type crusher of 200 tons per hour capacity, which reduced the stone from three foot cubes to three quarter inch particles. The limestone was then conveyed over the main road separating the quarry from the works on rubber belt conveyors to the works material store. The necessary shale was pumped from a washmill and storage mixer in the quarry to two 400 ton storage tanks adjacent to the millhouse. The limestone was fed with process water into a three compartment ball-mill by a pan feeder, while the shale slurry was added at the mill inlet by means of a spoon feeder. The quantities were regulated to give the required fineness and calcium carbonate content. The mill was driven by a 1,100 HP 3.3 KV motor.

The resulting kiln feed slurry was pumped by centrifugal pump through hydrocyclones which returned the coarse fraction to the mill for further grinding. The fines were pumped to four 400 ton doctor tanks where the slurry was blended by gravity feeding into a

Ewekoro from the air, November 1962

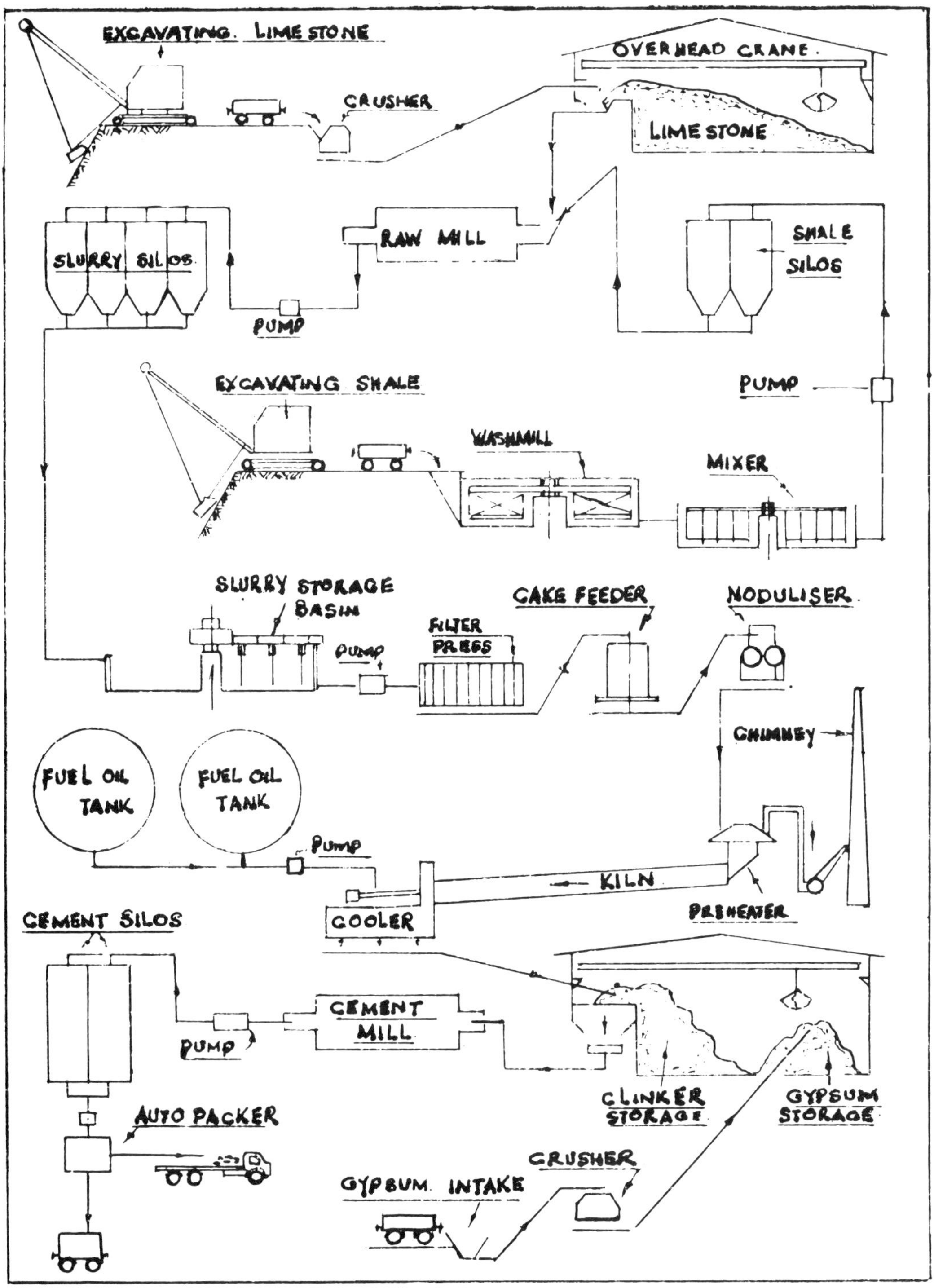

Times Press Lagos 2385/60

How cement was made at Ewekoro in the sixties

The kiln at Ewekoro

3,500 ton storage basin and kept homogeneously mixed by mechanical stirring and compressed air agitation. The slurry was pumped from the storage basin by centrifugal pump to filter presses which reduced the water content to about 20% leaving a filter cake, which was conveyed by rubber belt to the nodulising plant. This is what is called the semi-wet process.

The Davis pre-heater and exhaust gas duct to the chimney stack

The filter cake was then fed into variable speed drum feeders, which cut it into small lumps for feeding to the nodulising rolls. The nodules were conveyed via belt weighers to the preheater, where they received their initial contact with the burning process. The preheater removed the water content of the nodules and started to

The Davis pre-heater

decarbonate them before feeding them down to a 198 foot by 12 foot 6 inch diameter rotary kiln at a rated capacity of 25 tons per hour.

The pre-heater became known as the Davis pre-heater after the man who had developed it, Geoffrey Davis. He was the nephew of one of the early directors of APCM, AC, later Sir Charles, Davis. Geoffrey, as dedicated as his uncle in learning about the cement production process, became works managing director of APCM in 1953 after a career in APCM's northern empire. He had joined the company in 1929 as a technical assistant at Hope Works, become assistant manager in 1931 and manager in 1936. In 1943 he had moved to Hull to take charge of the northern area and had joined the board of G and T Earle. In 1945 he had moved to London as assistant to Ronnie Spier, works managing director and had taken over that position himself in 1953 when Spier retired. Like his father and uncle Geoffrey Davis believed in the application of scientific techniques to the manufacture of cement and he did much to promote scientific control and automation. The pre-heater was one of his developments and where the conditions and materials were right it worked well. Fortunately, Ewekoro was one of the sites where conditions and materials were right – fuel was expensive and the materials were low in sulphur and alkalies – and is in fact the only works where the Davis pre-heater continues to operate.

The kiln was fired by fuel oil supplied to the two 1,000 ton storage tanks by road tanker. (We have already seen that Willmott and Jones had recommended fuel oil as opposed to coal. Ewekoro was 400 miles from the coal fields of Enugu and oil proved to be the most economic source of energy.) The fuel oil was steam heated to 220 degrees Fahrenheit before being fired in the kiln by the return pressure system. A Thermax boiler, itself oil-fired, supplied the steam. Clinker from the kiln passed down a Fuller inclined grate cooler before being conveyed by belt to either the cement mill hoppers or into the material store.

Gypsum had to be imported and was brought in through the Apapa dockyards at Lagos before being transported by rail and unloaded by hand into a swing hammer crusher. It was then taken up to the material store by belt conveyor. The gypsum storage capacity was 2,000 tons.

The gypsum, clinker and limestone were all handled as required by a 10-ton Polysius overhead travelling crane and Polysius had also manufactured the cement mills. Reek and dust were extracted from the cement mills by a mechanically rapped dust plant containing six compartments of stockings. The cement from the grinding mills was pumped via Fuller-Kinyon pumps with compressed air to two cement storage silos, each with a capacity of 2,400 tons. From the silos, the cement was extracted on airslides to a chain bucket elevator delivering via a screen to the Fluxo packer hopper. This was a 12 spout rotary packer filling 50 kilo paper sacks at a rated capacity of 120 tons per hour. The filling and loading were done by packing teams on piecework. From the packer the filled sacks were conveyed either to retractable road loading conveyors or to a rail loading conveyor. There were also facilities for road bulk loading, but initially the works had to wait for suitable vehicles to be imported.

Power was supplied by overhead lines from the Electricity Corporation of Nigeria generators at Lagos to a substation alongside the works where it was transformed from 33KV to 3.3KV. The voltage was further reduced to 415v for the low tension supply. The management had also installed diesel-electric generators in reserve for turning the kiln and for essential services.

4

THE FIRST DECADE

Teething problems at take off

After the commissioning of the Ewekoro plant in December 1960, the sale of Elephant cement began in Nigeria on January 10th when Sir Adesoji Aderemi, the Governor of the Western Region, bought the first bag. The WNDC Report for the year ending March 1961 said,

> "The factory has not yet hit its stride as the machines are being run in. Moreover, modifications to a part of the machinery have been found to be necessary, and this means that production will be restricted for the next few months."

These were modifications to the Davis pre-heater which, though ultimately successful, was in its infancy and was therefore breaking new ground. The works was also short of filter capacity and a new filter press had to be added.

Two other problems were not so easy to cure. One concerned the frequent interruptions to the electricity supply. Power outages, as they were called, damaged equipment, particularly the kiln which ought to run continuously 24 hours a day except when shut down for maintenance or repairs. Each power outage necessitates rekindling the kiln and can cause damage to its lining. This problem continued to plague the works into the eighties. The other problem arose from the swampy nature of the Ewekoro quarry and the high water table level, liable to flooding in heavy rains. This sometimes slowed down the extraction of limestone. It also seems that the design of the plant did not make sufficient allowance for this. Part of

Tom Chesterfield showing Prime Minister Tafawa Balewa around Ewekoro works

the machinery, especially for cooling, was installed at underground levels, thus increasing the risk of flooding during heavy rains if, for example, there was a break in electricity supply or the pumps did not function properly.

The initial warm welcome accorded the brand name of Elephant Cement was soon followed by a period of serious marketing problems in 1961-62. This arose not only from competition from Nigercem which was older in the market, but much more from imported cement from Eastern Europe – Poland, Rumania and Yugoslavia. The cement was being dumped at prices less than the freight cost. Tom Chesterfield recalls that he was forced to protest to the Federal Minister of Commerce and Industries, threatening to close down the works unless something was done. The Minister said he would think about it. Chesterfield grew impatient and carried out his threat, closing down the works for ten days. The Minister was not pleased, but nevertheless eased the situation by imposing import controls, putting the importation of cement on licence, a move which was to have serious repercussions in the seventies.

WAPCO's main strategy, however, was to embark on aggressive marketing so as to get the brand name accepted as the favourite on the Nigerian market that up to then had known only the high quality brand, Burham, imported from APCM in the UK, and in recent years, Nigercem from the new plant at Nkalagu. By mid-1962,

Tom Chesterfield welcomes Earl Mountbatten after a game of polo at the Lagos Polo Club on 14th October 1964

it was being said that Elephant cement was the one local product, apart from Guinness of course, that Nigerians preferred to imported brands. Chesterfield had made sure that the Ewekoro production was of the highest quality and the advertising emphasised this. The Elephant symbol was chosen as it denoted strength. It had been used in the Burham advertisements also. Posters and cartoons in cinemas were used to promote the Elephant brand and distributors carefully selected. By September 1962, demand had outstripped supply. By minor modifications, production was raised by 17%, and proposals were mooted for a second kiln. Work started on this in 1963 and was commissioned by 1st April 1964. This raised the annual capacity from 200,000 to 400,000 tons. Almost immediately plans were made for additional filters and other anciliary equipment to raise production to 500,000 tons.

Industrial relations and training

The principal business of WAPCO to mine limestone and red shale and convert it into cement took place at Ewekoro works in a rural setting. In 1960 there was only a modest headquarters establishment located first in Niger House (UAC Headquarters), and later in the Barclays Bank Building in Marina. The UAC initially guided the new company in its marketing and public relations and it was UAC rules

Tom Chesterfield signing an agreement with the unions in 1961 (works manager, PR Breen, far left)

of personnel management and Staff Conditions of Service that served as models before the Company evolved its own rules and avoided further criticism that it was being run as a subsidiary of the UAC.

Manufacturing cement is a 24-hour, seven day week, intense activity. The works operated on a four shift basis. Being far from urban centres WAPCO, like the UAC in remote areas, had to set up residential estates for its senior and junior staff, and provide various municipal facilities: housing, water, electricity, transportation and, for the senior staff, also a golf course, tennis courts, supermarket, swimming pool and a clubhouse. In the 1960s senior meant expatriate and junior meant Nigerian. Inevitably, WAPCO took off from a colonial setting in which a predominantly expatriate staff had to recruit, train and supervise Nigerian staff.

Thus labour relations in the newly independent and decolonising society had its political and racial undertones, and called for human relations well beyond that expected in a normal industrial setting. The expatriate staff had to readjust colonialist views of "the native". They had to learn to train and guide staff, recognising their merits and abilities and advancing them within the heirarchy accordingly. They had to learn to work with African staff as partners and colleagues. They had to respond to the pressures of the Workers Union for a more stable basis of employment, with rules and due process for taking disciplinary action.

The industrial relations atmosphere in the 1960s was rather turbulent as it was characterised by strikes following workers'

An advertisement for Elephant cement in the sixties – getting the brand name established

agitation for improved conditions of service. This was compounded by the inexperience and lack of trade union education on the part of Union leaders. Both Union and Management were learning, sometimes with the intervention of the Industrial Arbitration Tribunal, to resolve their differences.

Besides the Workers Union, there were the Secretarial and Supervisory Staff Association (SSSA), and later also the Management Staff Association (MSA). The strain of adjustment to conditions in the industrial setting was probably hardest on Nigerian professionals joining the company in the 1960s and seeking to transcend the gap between the junior and the senior housing estates and in the allocation of benefits generally.

From the beginning, the emphasis was on training and staff welfare. As Tom Chesterfield said,

> "Although our financial investment is mainly in machinery, we are very conscious that these cannot function satisfactorily unless they are backed by a happy

Some employees who benefitted from WAPCO's training programme in the early sixties
Left: Sebastian Ogbu – dispenser in the works clinic
Right, top: A A Aninye – acting assistant works engineer
Right, bottom: S O Oyebanjo – assistant office supervisor

> and efficient team of men. To this end particular attention is given to the training and well-being of our employees."

When the construction of the works began in 1958 not a single local Nigerian had any experience of the cement industry or indeed of the skills and techniques associated with it. APCM took the opportunity to send several Nigerians to its locations in the UK for training. Others received training during the period of construction. Training was a continuous 'on the job' process, but was supplemented by sending selected personnel on booster courses at various Trade Centres and specialist training establishments. The company also set up its own training school. The emphasis in staff development was on experience, as in the cement industry a wide knowledge of many fields, chemical, mechanical, electrical and civil engineering, is essential.

The whole point of training was to produce cement efficiently and economically and also of the highest quality. Quality control was maintained by a fully equipped chemical and physical laboratory at the works under the supervision of a Nigerian, David Adeyemo BSc Tech (Hons), ARIC, AMI Chem E, who had qualified in the UK. To supplement the instruction in the training school and on outside courses, the company set up an apprenticeship scheme through which young Nigerians were trained as craftsmen and technicians. The four year course was designed to train each individual in more

Left: David Adeyemo – works chemist
Right, top: John Adeyinka – quarry supervisor
Right, bottom: Isaiah Babalola – charge-hand miller

than one skill. The training was both theoretical and practical alternating between the classroom and the factory.

Colin Booth, one of the expatriate technicians who served at Ewekoro for seven years from 1963 to 1970 and who was involved in developing the training scheme, can remember seeing many employees develop from fitters into supervisory roles. Saibu Adebakin, currently assistant chief engineer at Ewekoro, was one who benefitted from the training programme. He was recruited while the works was being constructed, having worked on his cocoa and kolanut farm. For him – as for many others – the works was good news bringing more secure and better paid employment. He began as a labourer and after the commissioning applied to join the apprentice scheme. He was taken on as an apprentice fitter-machinist and as well as being trained at Ewekoro he was sent on courses at the UAC training school at Burutu. From apprentice he progressed to workshop charge hand and then up the ladder to shift foreman, to shift supervisor, to assistant section engineer, to section engineer and to assistant chief engineer.

Duro Ibagun was another to benefit. He can remember the geological survey in 1957 though he had no idea at the time that it was for a cement factory or that he would spend most of his life working in it. He joined the works as a labourer in the mill house in May 1964 and was paid 7.5 kobo an hour (about 175 naira a year). In 1966 he transferred to working on the kiln and stayed there until he returned to the mill house as foreman in 1978. He then went back to the kiln as production supervisor and from there he progressed to assistant manager production department, to chief production manager, and to production services manager.

The company also instituted a functional literacy programme and paid £5 to those who completed the course. Many, such as bricklayers, had been able to complete trade tests once they had learned to cope with the written parts of the exams. At the other end of the scale, training for supervisory and management positions was instituted with the guidance of the chief training officer of APCM in the UK.

By 1965 the Ewekoro works was well established and WAPCO seemed to have negotiated the first tricky years of a new venture. Adisa Adeleye, the journalist, who was not always complimentary about businessmen and politicians, wrote in glowing terms of the success of the factory and the company, using such phrases as "worthy venture", "picturesque scene", "well integrated factory", "towering and shining silos". He also remarked on the significance of local manufacture and the saving of foreign exchange. Only gypsum needed to be imported through France,

> "The importance of this is that the industry saves for the country a lot of foreign exchange and by its consumption of local products helps expansion in the industries supplying these products."

He also noted with approval that all eight of the senior supervisors in the factory were Nigerians. He went on to address the question of why locally produced cement was not cheaper than imported cement. Cement users and distributors had been complaining of the recent rise in prices and telling government they were now paying more than when they had to rely exclusively on imports. Government responded by issuing warnings to the cement manufacturers. As WAPCO pointed out part of the reason was the government's fault. The explanation lay in the hidden costs. WAPCO had to pay sixpence (5 kobo) royalty to government on every ton of limestone produced, ninepence (7.5 kobo) excise duty on every ton of cement, and import duty of five per cent on spare parts. There was also depreciation on plant and the high cost of imported skilled labour. The company was also able to point to harbour dues on the shipment of oil and the high cost and unreliability of the electricity supply. Furthermore WAPCO was a leader in providing a substantial amount of social services to its employees which in industrialised nations would have been provided by the state through taxation systems.

Political problems

The marketing strategy of WAPCO had to take into consideration the political rivalries and divisions which, beginning from May 1962, destabilised the Western Region Government, its sponsor and a major shareholder through the WNDC. Relations between Chief Obafemi Awolowo, the Action Group Leader, now Leader of the Opposition in the Federal Legislature, and Chief S L A Akintola, the Premier of the Region, deteriorated and the Action Group was split into two factions. The larger section under Chief Awolowo signed a document declaring loss of confidence in the Premier and asking the Governor to remove him from Office. The Premier went to Court to seek a Declaration that only the House of Assembly meeting in session could pass a vote capable of removing him from office. An attempt to hold a meeting of the House was twice disrupted by the smaller faction in the House. The Prime Minister, Alhaji Tafawa Balewa, in the Federal legislature declared that law and order had broken down in the Western Region, dissolved the House of Assembly, suspended the Government and appointed a Federal Commissioner sympathetic to the Akintola faction to run the Government of the Western Region for six months. Within that period arrangements were made to carve the non-Yoruba parts of the Region into a fourth Region called the mid-West Region. A

Chief S L Akintola, Premier of the Western Region, addressing a public meeting. Seated next to Chief Awolowo is Oba Adesoji Aderemi, Governor of the Western Region, who commissioned Ewekoro works in 1960

Alan Taylor, chairman and managing director from 1966 until 1971

judicial enquiry was ordered into the finances of the various statutory corporations, including the Western Nigeria Development Corporation. Then Awolowo was arrested, later tried for treasonable felony and imprisoned. The instability of the Western Region soon spread, leading to widespread disorder by 1965 and was a major factor in the military coup of January 1966.

WAPCO reacted to these events by making and selling cement and staying aloof from politics. The crisis in the Action Group was nothing to do with the company. If that crisis and the later disturbances in the Region sometimes disrupted distribution, it was merely an incentive to seek markets further afield and make Elephant Cement known and welcome throughout the country. Alan Taylor succeeded Chesterfield as chairman and managing director in 1965. He had joined APCM in 1952 as a sales trainee with the Cement Marketing Company (CMC) and had transferred to the Overseas Investment Department under Arthur Poole and John Milne in 1962. He continued to expand the sales effort and set up depots throughout the country at Ibadan, Ilorin, Kaduna, Minna and Jos. As mentioned before, Nigerian railways built sidings at the Ewekoro works to facilitate evacuation and distribution of cement. The policy was to take the cement to the consumers as much as possible by rail, rather than rely completely on distributors. Thus, rather than suffer from the political crisis, WAPCO was able to

Sir Mobolaji Bank-Anthony, a well-known Nigerian businessman and WAPCO distributor (far left) with the wife of Alan Taylor (second from right).

Major General Aiguyi-Ironsi – the new head of state in January 1966, with Vice Admiral Wey and Inspector General of Police, Kam Salem

expand production and sales. In fact demand continued to run well ahead of supply, and plans began to be discussed for further expansion at the works.

The Civil War years

As the expansion plans were being put into action, the political scene in Nigeria darkened further. In January 1966 Tafawa Balewa the Prime Minister, Okotie-Eboh the Federal Minister of Finance, Akintola the Premier of the Western Region, Ahmadu Bello the Premier of the Northern Region, and several senior Army Officers in Lagos and Kaduna were assassinated in an attempted military coup. The Commander in Chief of the Army, General Aguiyi-Ironsi, who rallied the loyal troops to contain the mutiny, was then invited to take control of the Government. In July 1966 there was a counter coup, led by officers of Northern Region which brought the then Lt. Colonel Yakubu Gowon to power. The Military Governor of the Eastern Region, then Colonel Odumegwu Ojukwu, refused to

Ironsi was soon replaced by Lt. Col Yakubu Gowon. On his left is Maj. Gen. David Ejoor, Chief of Army Staff, and on his right Vice Admiral A Wey, Chief of Naval Staff

recognise Gowon as Head of State and, in view of widespread killing of Igbo army officers and civilians in the Northern Region, threatened to secede from the Federation. To meet the crisis, Gowon split the Northern Region into six states; the Eastern Region into three states – the Igbo East Central State, the South-Eastern State and the Rivers State. Lagos was constituted into a separate state; the mid-West Region became the Mid-West State, and the rest of the Western Region became the Western State, controlling WNDC and the shares in WAPCO. Awolowo, who had been released from jail in August 1966, became Vice-chairman of the Federal Executive Council that controlled operations during the civil war which broke out in July 1967.

For two-and-a-half years the country suffered a divisive, damaging, bloody and expensive civil war. Apart from the general destabilisation, the main effect on Ewekoro was the disruption caused by a number of their staff from the East leaving to return home. In terms of competition from other cement factories, the war meant cessation of production at Nkalagu in the East. The factory at Sokoto in the north west came stutteringly into production but the new one at Ukpilla (APCM had surveyed that area but decided against it in favour of Ewekoro) was not yet commissioned. While the war lasted, WAPCO could sell as much cement as it could make and pursued plans to increase its output.

During this period, the Government took measures to protect the Nigerian balance of payments, stimulate domestic output and help the country mobilize its resources fully for the civil war effort.

The elusive search for peace at the Aburi conference – Gowon greets Col. Ojukwu with Col. Adebayo, Military Governor of Western Region, looking on at a meeting hosted by military leaders in Ghana. The failure of the meeting led to civil war

Towards this end the Government limited the repatriation of profits and dividends, introduced a rationing scheme for foreign exchange payments, limited the proportion of overseas allowances to public officers serving abroad and clamped down on non-development use of foreign exchange.

Fiscal measures were introduced to limit and redirect consumer expenditure on imports and rechannel it to more productive uses. Taxes were imposed on the excess profits over total capital allowances for companies enjoying pioneer status. At the same time, taxes were imposed on companies' turnover where no profits were declared. The duties on consumer goods were raised under the Customs Tariff (Duties and Exemption Act).

Thus it is understandable that local cement production and imports dropped in volumes during the war period. While the drop in local production output was due to the closure of the cement factory at Nkalagu, the reduction in cement import flow was due to some of the fiscal measures taken as explained above. It is not surprising therefore that the market share of WAPCO became very high during this period.

The Federal Government decided that strategic factories should be protected against sabotage and a platoon of soldiers was billeted at Ewekoro. Nevertheless, demand for cement continued above the

Oil played a crucial part in the Civil War, as indeed it has in Nigeria's development of the last 30 years. Here the oil refinery at Okrika, near Port Harcourt, after capture by the Federal troops during the Civil War

level of supply and in July 1966 Alan Taylor wrote in the house magazine *Elephant News*,

> "With an increasing volume of cement being sent to the Mid West and a surprising buoyancy in our traditional markets, despatches in recent months have been brisk and so far are on target for the year. Production has been satisfactory, in fact creditable considering the electricity restrictions which, I regret to say, have again influenced our performance in the last quarter, reducing output and increasing costs."

He went on to say that the decision had been made to invest N1.3 million to consolidate production. He highlighted the Second Mainland Bridge, now known as the Eko Bridge, joining Lagos island to the mainland as an example not only of an expanding economy, but also of the quality of Elephant cement which made it the choice of the contractor, Julius Berger. Between November 1965 and May 1968 39,256 tons of Elephant cement was supplied for the bridge and when it was opened in July 1968 it was the largest prestressed concrete bridge in Africa.

5

EXPLOSIVE DEMAND FOR CEMENT

THE SUPPLY of cement in Nigeria had been struggling to keep pace with demand. As the civil war ended demand grew even faster. Not only were there many damaged buildings and bridges to be reconstructed but new projects were also planned. For 30 months, from July 1967 until January 1970, civil war had raged and any development project not related to the war had been put in abeyance. Now, as well as the repair work, the government was determined to embark on new projects such as airports, dams, harbours, expressways and overhead bridges. Both the Federal and State Governments invested funds in construction companies. In particular the Federal Government needed to build army barracks as the army had grown from a virtually ceremonial strength of 8,000 in 1967 to a fighting force of 200,000 in 1970. This large body could not be demobilised instantly without serious social and security implications, and barracks were urgently required to house them in order to release public institutions for their regular use and ease housing shortage for the civilian population.

With this urgent need for cement the Federal Government granted £1.4 million for the rehabilitation of the five cement factories in the country – WAPCO amongst them – which strove to produce as much as possible. All these could be expected to produce only 50% of the country's needs. As part of the drive to produce more cement, WAPCO in 1972, with Richard Cheney now at the helm, installed a third kiln at Ewekoro bringing production capacity up to a realistic level of 750,000 tonnes a year. Installation was completed during 1973 by which time the oil price had quintupled and demand for cement was set to rocket in the ensuing construction boom in Nigeria. The government now felt it had as much money as

The exploding demand for cement.

Shell flats at First Avenue, Ikoyi

Offices for the British pharmaceutical company, Glaxo

Book Shop House, Broad Street, Lagos

A factory for Metal Box

it could spend on development projects; its ministries became over-confident, and cement import licences were issued to all and sundry without control and without regard to the ability of the recipients to use them or control the ensuing influx.

The result was that no less than 279 ships carrying cement were 'parked' outside the docks at Lagos by October 1975. What had caused this massive bottle-neck was the loss of control in government ordering, so that between May 1974 and June 1975 Nigeria ordered 21 million tonnes, 16 million of them ordered by the Ministry of Defence. The total value of cement ordered was put at N650 million, with the Ministry of Defence orders valued at N609 million. By October 1975 there were 3 million tons in ships waiting to berth, 16 million on the way and a further 5 million ordered for development projects. As Lade Bonuola reported in the *Sunday Times,*

> "The total yearly discharge capacity of cement at the wharf is 1.4 million tons. But at the time orders were being placed, it was envisaged 1.6 million tons would be discharged every month without taking the physical constraints at the port into consideration."

The *Tribune* added its voice with a thundering leader,

> "Whoever was responsible for the fantastic order of 16 million tons of cement for the Ministry of Defence committed an unforgivable crime against this nation. And he should not only be exposed, he should be given the severest possible punishment. ... The simple arithmetic of the whole dirty business is alarmingly distressing. According to the Chief of Staff, Supreme Headquarters, the rate of discharge of the cement cargo has increased from 2,000 tons a day to over 10,000 tons. But this statement gives us the impression that the Federal Government is not fully aware of the enormity of the crisis.
>
> "We have ordered 16 million tons of cement. Four have already arrived of which 2 million are in our waters awaiting berths. At the rate of 10,000 tons a day it would take 200 days to discharge those already in our harbour. That takes us to May 19, 1976. The remaining 12 million would be discharged at the present rate in 1,200 days. That takes us almost into 1980! It would mean that hundreds of ships would be lying in Lagos/Apapa harbour for four years!"

To try and cope with Nigerian demand hundreds of thousands of tonnes were shipped from APCM's new factory at Northfleet, Kent, on the site of the old Bevans works

The new government, which had replaced General Gowon's administration in July 1975, tried to get to grips with the situation but the extent of the excessive ordering and the onerous nature of the contracts, including provisions for delivery within 12 months, payment upon shipment and heavy demurrage charges, were not immediately apparent. When the scale of the scandal became known the government took firm steps to re-negotiate contracts and slow down the shipments. Re-negotiation was certainly necessary for it

Richard Cheney (left), chairman and managing director of WAPCO in the early seventies with Ade Odeniyi, Sales Director, and Dr Gordon Marshall, deputy chairman of Blue Circle Industries in the eighties

was calculated that shipment of 20 million tons in one year – the National Plan called for that amount over five years – could cost the government 1.6 billion US dollars in subsidies and demurrage charges.

We shall see in chapter seven how the new government tried to cope with the oil boom and its effect on Nigeria. In the meantime how did WAPCO react to this cement shortage and exploding demand?

One of the reactions of WAPCO to the explosive demand for cement in the early seventies was to change its method of distribution and charging. Before 1974 cement was sold on what was known as the "Authority to Collect", popularly known as ATC, basis. Under this system, customers, having paid for their cement, were given the relevant documents to collect it from the factory. By 1974 it was felt essential to change this system which had become totally confused. It was impossible to determine how many vehicles would be at the factory at any time of the day. Dr Gordon Marshall, deputy managing director of Blue Circle Industries in the eighties, can remember constant queues of lorries at the Ewekoro works with drivers servicing their engines and changing clutches. Moreover it was difficult to establish the destinations of these vehicles. What was happening in practice was that the ATC documents were being

To try and beat the blockage in the ports, APCM shipped cement in bulk to Flour Mills in Apapa –

- loading bulk cement tanker

- cement silos

● reception of imported bulk cement

passed from hand to hand with each transaction attracting a mark-up or commission.

Tony Jackson, who took over as chairman and managing director of WAPCO at the end of 1973, immediately abandoned the ATC system and told a press conference in Lagos in January 1974 that end users in Lagos were paying 56 naira per ton against an ex-factory price of 26 naira. The difference was the cost of "distribution". He hoped that by taking over this distribution its cost could be reduced. WAPCO intended to create distribution zones within which sub-distributors would be appointed. The sub-distributors would buy cement from the zonal warehouses at prices controlled by the company. Special allocations would be made for projects of national importance and the company would deliver direct to the contractors. He stressed that distributors in future would have to be organisations with a disciplined market approach and possess proper warehouses and offices.

By the middle of 1974, when demand in the rest of the world began to ease after the oil price explosion of the previous winter, it was only just beginning its steep ascent to the peak of October 1975. Nigeria, after all, was a beneficiary from the price rise. However, inflationary pressures were building up and in August 1974 WAPCO raised its prices ten per cent. Nevertheless, the real problem was not price but supply.

Flour Mills

Whilst, as we shall see later, WAPCO was embarking on a further major expansion, Jackson recognised that immediate short term measures were necessary, and felt that any initiative possible should come from WAPCO, the leading Nigerian producer, and APCM, its technical partner as a recognised world leader in cement issues. In order to facilitate the construction of the Kainji Dam in the sixties the contractors had installed purpose-built silos for the reception of the special quality bulk cement required. These silos at the cessation of the contract had been taken over by Flour Mills of Nigeria for the importation of bulk grain. Before the "cement armada" Jackson approached Flour Mills with a view to acquiring these silos and establishing an import bulk terminal in order to meet the shortfall prior to its next expansion.

Flour Mills, which was under the dynamic direction of George Coumantaros, declined to dispose of the silos but, recognising the commercial opportunities, sought to become involved in the scheme proposed by Jackson. Flour Mills' participation was negotiated on the basis of APCM supplying the bulk cement from its Northfleet plant which would be shipped in the vessels, suitably converted, of Southern Star shipping company owned by Coumantaros, to be received at a depot with a packing plant facility added at the expense of Flour Mills. All the operations were to be managed, and the cement distributed, by WAPCO. Facilities at a cost of 12 million naira were being completed and the first two bulk ships at sea, when the Nigerian government became suspicious of the operation and determined to nationalise it.

Following the negotiations between Coumantaros and Jackson, the government subsequently agreed (having satisfied itself that the offshore supply contracts had in fact been negotiated on terms entirely satisfactory to Nigerian interests) to WAPCO and Flour Mills continuing to manage the operations under the supervision of the National Supply Company who subsequently confirmed the shipping and cement supply contracts. Moreover, they organised a revolving letter of credit which greatly facilitated the operations. The great advantage that this scheme had over the generality of cement imports was the ability of the special ships to sail without delay through the moored armada to Flour Mills' dedicated wharf in Apapa and to discharge its 30,000 tonne cargoes in a matter of hours. This economic supply arrangement resulted in cement prices being considerably below those of other imports and through considerate and disciplined distribution arrangements ensured that vital government projects were built within a reasonable time and sensitive private demand was at last tackled. These arrangements continued until the change to civilian rule in 1979 but have continued to a lesser extent under the direction of Flour Mills, WAPCO by then having concluded its expansion programme.

6

COMMISSIONING OF SHAGAMU

The planning of another factory

WAPCO reacted in a logical and businesslike way to the excessive demand for cement by commissioning another factory to support its factory at Ewekoro. Given the demand for cement there were a number of organisations wishing to increase cement capacity. Apart from WAPCO's own ambitions, both the Western State Government and the Federal Government wanted to construct plants utilising the

Tony Jackson, chairman and managing director of WAPCO in the mid seventies, welcomes Brigadier Oluwole Rotimi, Military Governor of the Western State, to the foundation ceremony for the Shagamu works in early 1975

Tony Jackson introduces Norman Mullins, chairman of APCM, to Rotimi at the ceremony

raw materials available in the West. Following a series of complex negotiations it was finally agreed that a new factory be built based on the raw materials available at Shagamu by WAPCO with the equity of the company adjusted to facilitate a 20% participation by the Federal Military Government. This infusion was complemented by a substantial Government loan of N24 million. The company's then bankers, Barclays and Standard, were also helpful in providing loans in the form of overdraft facilities to meet the shortfall.

The chosen site was to be Shagamu, not far from Ewekoro, and utilising the same belt of limestone discovered by the APCM research department geologists in the late fifties. In helping with the funding the Federal Government took a stake in the company. The new shareholdings in the company would be – APCM 40%, the Western State Government 32%, UAC 8% and the Federal Military Government 20%.

By early 1975 the proposed site had been cleared and a foundation ceremony took place at which the Military Governor of Western State, Brigadier Oluwole Rotimi, unveiled a plaque of a Blue Elephant to symbolise the construction. The ceremony was also attended by several Obas and Chiefs, the British High Commissioner and State Commissioners. From the UK came the chairman of APCM, Norman Mullins, accompanied by John Milne and David Baird. Tony Jackson was then managing director of WAPCO. He had

Tony Jackson between Rotimi and Sir Martin LeQuesne, the British High Commissioner, with the Akarigbo on the far left

worked in APCM since 1957 after five years active service in the British Army. After various appointments within the UK he had joined the Overseas Investment Department in 1964 and under Arthur Poole had been involved in both the financial and commercial supervision of APCM's overseas investments and in the investigation and establishment of new ventures. After a spell in Australia as assistant managing director at APCM (Australia) he was seconded to WAPCO as chairman and managing director in 1973. He recalls that,

> "At that time perhaps the additional capacity was expected to be installed at or near Ewekoro. However, following discussions with the Western State Government we were persuaded, in the interest of spreading employment throughout the State, to consider alternative locations. We asked our technical partners, The Associated Portland Cement Manufacturers, to put their geologists to work and they came up with the possibility of basing a new works here at Shagamu."

Jackson went on to make the point that the siting of a cement works was influenced by other factors as well as the availability of raw materials. Closeness to major markets, the proximity of a suitable workforce, good road and rail communications and reliable water and electricity supply all played their part. The location at Shagamu, in preference to Ilaro or Owo or next door to Ewekoro, satisfied all the requirements except rail communication, but it was conveniently situated between Ibadan and Lagos and a major trunk road joining the two to the main route to the East and Benin was under construction with a link road to Shagamu already at the planning stage.

Jackson also acknowledged the co-operation the company had received from the Akarigbo of Ijebu-Remo. The building of a large cement works would not be to everyone's liking, but he hoped the enhancement of employment opportunities would more than offset any inconvenience or dust and noise pollution. Both Jackson and the Western State had written to the Akarigbo in 1974 telling him of the plan to build the factory and he in turn had informed the local Obas so that they could tell the villagers. WAPCO then invited the Akarigbo and Obas to talks in March 1974 so that the principle of compensation could be agreed.

The government was supposed to negotiate purchase of the land with the local people but had not carried out this task. This upset the landowners who were threatening the geologists. Jackson therefore visited the Akarigbo – taking his wife, Jo, with him for protection! – and discovered that the people were indeed upset as they thronged round his car. He and Jo went in to see the Akarigbo and with the aid of maps showed him exactly what was planned. The Akarigbo, quickly aware of the benefits in terms of employment, told his people that he was fully in favour and a potentially explosive situation was defused.

The Oba at Iperu can remember that there was something of a tug-of-war between those who did not want to give up their land and those who saw that the factory would bring greater prosperity to the area. The plant was expected to employ 600 people directly and many others indirectly. The Oba persuaded those that did not want to sell their land to compromise because they would face compulsory purchase if they refused. A figure of N51 per acre was agreed to be paid by the Western State Government.

Jackson was able to say at the foundation ceremony,

> "For our part we recognise some of the unattractive features of heavy industry and are doing our best to overcome them or at least minimise their effect. Our design will incorporate the latest equipment on dust suppression and embrace all the most modern techniques. Also some of you may have wondered why

> it is that we are tucked away today in this bush clearing. The fact is that we are conscious of both noise and appearance, accordingly therefore we are leaving a barrier of bush some 200 metres wide between the factory site and the road. This will not only help deaden noise but also avoid the inevitably somewhat unsightly industrial activity from the view of the passing public."

Brigadier Rotimi in his speech explained why the government were so keen to help finance another cement factory,

> "A study commissioned recently revealed that cement consumption forecast over the next ten years, will increase at the rate of 9% per annum. This was based on a 7% per annum growth of the Gross Domestic Product for the country and a population growth rate of 3% per annum. It is now well known that the Nigerian economy has been growing at a rate of more than 9% per annum and if the present trend continues, the growth rate during the Third National Development Plan period will be prodigious. I have given this brief statistical background in order to show that even the projected demand of 2,750,000 tons of cement by 1977 against a

Chief Awolowo and his wife were also present at the ceremony

Norman Mullins greets the Akarigbo

total local production of 1,700,000 tons would appear to be an underestimation. It was to ensure that construction works are not seriously disturbed in the private and public sectors that my government included the establishment of a second cement factory in the current National Development Plan. ... The Shagamu factory is being established at an estimated cost of N62,970,000. In view of the strategic position of the cement industry in our country's economic development, the Federal Government has responded admirably to the request for its participation and has accordingly provided a sum of N8 million as equity capital and N24 million as loan for the project."

The construction of the works

The Shagamu factory was to be significantly different from the one at Ewekoro. When the Ewekoro factory had been planned Nigeria's large oil reserves had not been proved, unemployment was high and cement demand much smaller. The factory was therefore designed on fuel efficient if not labour saving lines. The Shagamu factory was designed on a much larger scale incorporating the latest techniques

Aerial shot of
Ewekoro

Dawn over Ewekoro

Night at Ewekoro

Ron Burnett and David Adeyemo at Ewekoro in October 1977

Control Room at Shagamu, April 1979

Kilns at Shagamu

Mr and Mrs Ron Burnett's farewell present

A PPP Distributor – Mr Deji Ariyibe outside his depot in Ibadan

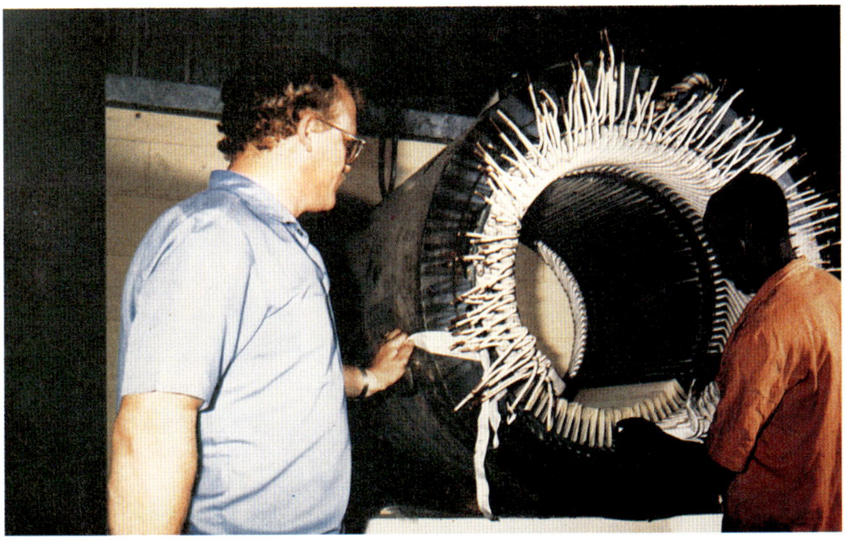

PER – the repairing of high tension motors

Residential complex in Lagos, decorated with Sandtex textures

Sir John Milne presenting a Blue Circle scholarship during a visit to the Akarigbo's palace in 1989

Akin Ogunleye, managing director, with his wife Yinka

Arriving for the commissioning of new equipment at Shagamu quarry in 1989. From left to right: Dr Ransom Murray; Chief Ogunlaja, the Lisa of Ofin, Shagamu; Johnson Otubaga, the general works manager; Akin Ogunleye, chief executive; the Alaperu of Iperu, Oba Ogunfowora; Chief Awofala, the Losi of Ofin, Shagamu; and Oba Ogunsowo, the Ewusi of Makun

The chairman of WAPCO, A A Abidogun

The model of the new Elephant Cement House at Ikeja

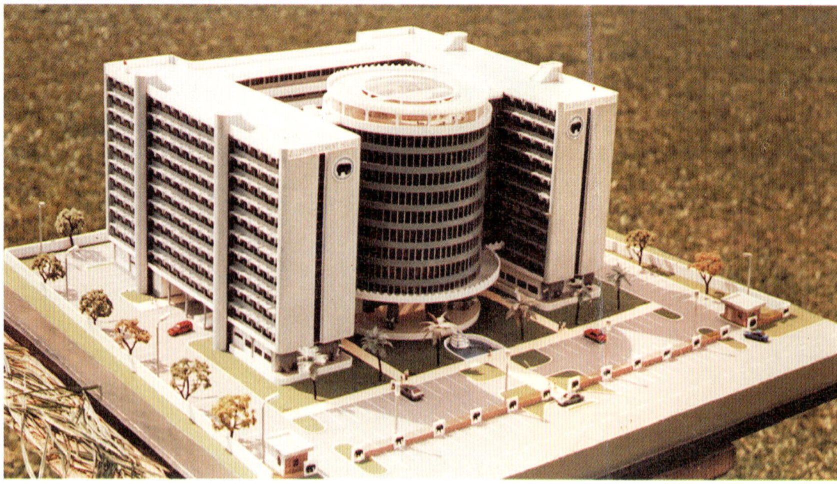

An RH75C back-hoe hydraulic excavator in operation at Shagamu in 1989

The Blue Elephant

One of the benefits of expatriate staff arose through the energies and skills of accompanying wives, who were not allowed to have gainful employment. Many of them became deeply involved in charitable and supportive activities. Here is Mrs Jo Jackson at Ife Oluwa clinic in Maroko for which she organised funding of the wet process. A guaranteed output of 600,000 tonnes was to be produced from two kilns 500 feet long and 14 feet 6 inches in diameter. Each kiln would have an output of 1,000 tonnes a day.

Initially, the 300 acre site had to be plotted and mapped and the owners identified so that they could be compensated. 160 plots were discovered on which grew 60,000 kola trees, 4,000 palms and 29,000 cassava plants as well as pineapples, oranges, bananas and yams. This work was completed by the end of 1974 and over the following two years the same exercise was carried out over the area required for the quarry. By the end of that time 2,700 acres had been covered.

It was only on 14 November 1974 that the Western State Government announced formally in their Gazette the acquisition of the land, effective from 28 November. WAPCO placed the first contract, that of site clearance, by 2 December and work started within a week. However, the consultants from APCM had been working for a year drilling boreholes all over the site and sending back samples to APCM's research department at Greenhithe. These had been analysed so that the engineers could decide on the type of plant required and the chemists the type of mix of materials and the geologists recommend the best way of working the quarry.

While Ewekoro was established in a rural area, Shagamu works is on the outskirts of a growing urban area, and has been able to take

The ground is cleared

Construction proceeds

The kilns are in place

advantage of both the urban and the rural settings. Accommodation is provided for some of the junior members of staff and trainee apprentices within Shagamu township, while a housing estate for the senior staff along the same lines as the one at Ewekoro is located some 2 km from the works further away from the town. In addition to the normal facilitites, a school was also provided for the primary education of the children of the estate and others in the community.

The clinker gantry and 5,000 tonne silos

Indeed, the quality of the school, like that of the staff clinic and other welfare facilities, were soon to become part of the WAPCO attraction, bringing considerable advantages to the whole community and offsetting whatever inconveniences might be associated with the location of the factory so close to town.

Preparation for the commissioning of Shagamu

Recruitment and training of personnel at the new Shagamu factory began well in advance of its commissioning in 1978. The first recruits began their training at Ewekoro as early as 1975. Throughout 1976 and 1977 the training department devised and arranged a multitude of courses using the most up-to-date training aids specifically designed for the new factory. Every single recruit in the 800 staff for the factory received some pattern of formal lecture room training. The training at Shagamu was aided by the fact that it already had a plant at Ewekoro and a residential training school.

Thirty thousand tonnes of plant and machinery had been brought from all over the world through the docks at Apapa with some individual items weighing over 79 tonnes. Over 50 wide loads had been transported on Sundays under police escort to avoid weekday traffic congestion. Over 2,000 people were involved in the construction. In the quarry over seven million tonnes of overburden were removed to expose the limestone.

The works, virtually complete – ample room left for future expansion

How cement was made at Shagamu in the seventies

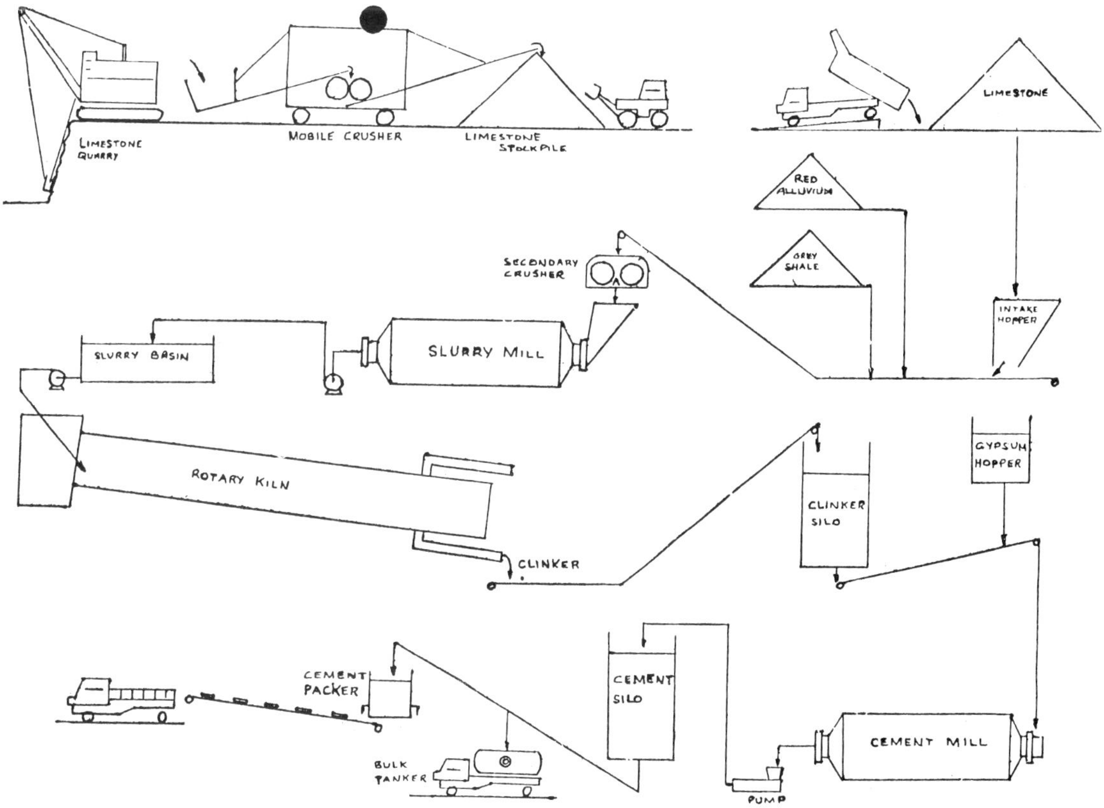

SHAGAMU WORKS - SCHEMATIC FLOW SHEET

Chief Olu Akinkugbe, the first Nigerian chairman of WAPCO, with his wife, Janet

By the second half of 1977 the factory was preparing itself for full production. On 22 September the first raw mill produced the first raw material feed or slurry. The company invited the Akarigbo of Ijebu Remo to witness this in gratitude for all his hard work in assimilating the plant into his area of influence. This was symbolised by naming the mobile crushing plant after him. The gradual process of initial ignition of the first kiln was started in November and the first cement clinker was produced in early December. By the turn of the year complete cement production was possible and after the necessary 28 days of testing the quality as concrete, the first cement delivery was made ceremoniously to the Governor of Ogun State, Lt. Colonel Ayodele Balogun at Abeokuta on 14 February by Chief Olu Akinkugbe, WAPCO's new Chairman.

Akinkugbe had trained as a pharmacist and in 1952 started his own business, manufacturing, wholesaling and retailing pharmaceuticals. Over the years he had been associated with chambers of commerce and had been both president of the Ibadan Chamber of Commerce and vice president of the National Chamber of Commerce. He had also been invited on to the boards of Barclays Bank (Nigeria) and the trading company, Paterson Zochonis, as well as becoming chairman of the Nigerian Industrial Development Bank. He served on a number of government commissions, notably the Morgan Wages Commission of 1963/4. He had also been chairman of the Public Accounts Committee of Western Region. The shareholding arm of the Western States governments, now called Odu'a had asked him to be one of their representatives on the board of WAPCO and in 1975 the WAPCO board invited him to be their first Nigerian chairman.

His view of WAPCO was that it was a company unique in Nigeria in that its product was considered superior to imported competitive products. The company's concentration on quality was outstanding. The company was also highly respected for its forward thinking policy in staff development. Not only was he, a Nigerian, chairman but there were many other Nigerians in senior positions. Its managing director, Tony Jackson, related to the aspirations of the country, and had voluntarily relinquished the chairmanship in favour of a Nigerian.

Having learned from their experience at Ewekoro, WAPCO had designed and built a simple and rugged operation at Shagamu without trivial embellishments, using tried and tested components from markets throughout the world. The processing plant consisted of only six basic units. Four of the units were powered by exactly similar 3,000 horse power motors and gear boxes. Wherever possible components had been standardised.

As the efficient operation of cement rotary kilns demanded the continuous supply of electricity WAPCO installed 5.4 Mw of diesel generation to power the sensitive kiln plants during periods of power outage. The total power demand was 14 Mw and the company had long negotiations with NEPA on the subject of constant and adequate electricity supply.

The opening ceremony

Three years after the laying of the foundation stone, on 5 May 1978, the dignitaries all returned to Shagamu for an initial opening ceremony. It had been a great achievement with the factory completed virtually on time and within the planned budget. All involved could feel proud of themselves and were congratulated by John Milne, now managing director of Blue Circle Industries Limited (APCM was renamed Blue Circle Industries in 1978), who had again flown out from London. He talked of continuity which has always been (and remains) a feature of Blue Circle's and WAPCO's involvement in Nigeria.

> "Looking at the audience today and seeing the guest list I cannot but help reflect on this theme of continuity. I see many faces and names well known for their contributions over the years to the advancement of this company. Some of them like Tom Chesterfield, the first

> chairman, returning to Nigeria after 14 years, made their mark in the management field. Others, such as Chief Awolowo and Chief Sowole played an important part in the initial conception of WAPCO way back in the 1950s."

Chief Olu Akinkugbe, the chairman of WAPCO, was also congratulatory and he took the opportunity to warn the government about excessive taxation of companies,

> "No one can be expected to invest his personal funds in a long-term venture in the face of government equivocation on the private enterprise issue. But a firm commitment to private enterprise is only part of the story. In addition the government must permit through appropriate fiscal and regulatory policies a fair return to be earned on private industrial investments. I find the present financial situation of the country a bit unsettling for as government finds its revenues falling short of it expenditure plans, there arises the temptation to tax readily visible sources. In the early years of the nation's independence, government found it easy to heavily tax farmers through the various produce marketing boards. The result was to discourage agricultural investment in cash crops and the detrimental effects of this taxation policy are still apparent to this day. We must ensure that the same thing does not happen to industrial enterprises. For to tax productive enterprises more than their fair share would be to kill the goose that laid the golden egg."

Akinkugbe also made the point that industrial success would not come to Nigeria just because people wanted it,

> "I hope we are not only witnessing the beginning of a long and prosperous life for this cement factory but the beginning of a new and prosperous era of sound industrial investment in the country as well. We have the raw materials, we have investment capital and we have increasing numbers of well trained and motivated personnel. The task of developing industrial activities fed by our own raw materials will not be easy. As this factory so amply demonstrates, the job can be done. But it needed substantial financial input from the Federal Government, the former Western State Government and

> now Odu'a Investment Corporation, the NIDB, several Nigerian commercial banks including ACB, Barclays, IBWA, National, Savannah, Standard and UBA in addition to credit arrangements by our technical partners – APCM."

The Honourable Federal Commissioner for Industries, Dr RA Adeleye, who performed the opening ceremony, added his warnings and hopes to the congratulations in the air. He outlined the need for such a new factory by explaining the exploding demand for cement.

> "Various attempts have been made to estimate the demand for cement in Nigeria during the next few years. For example, this year it is estimated that 5.4 million tonnes will be consumed. Of this figure only 2.4 million tonnes will be produced locally. This means that about 3 million tonnes, ie more than 50% of the estimated total demand will have to be met by importation. Looking further ahead, it is estimated that by 1980 the total demand for cement will be about 6.3 million tonnes. This assumes that all the new cement works (Shagamu, Ashaka and Benue) will be in full production and the planned expanded capacities of the existing plants – Nkalagu, Ukpilla, Calabar and Sokoto would have been achieved. By 1980 we will still have to import about 2.3 million tonnes."

Dr Adeleye went on to congratulate WAPCO on its decision to put in hand expansion plans even before the factory was fully operational and he highlighted the fulfilment of the government strategy of the dispersal of industry,

> "The dispersal of industries to urban and rural areas helps to open up such places to even development and provide employment to more Nigerians. The Shagamu project achieves this ideal. Starting as a cement town, Shagamu is bound to achieve accelerated development socially and economically."

But he gave a severe warning on a major cause of concern to the company, the profiteering of some distributors. He said that everyone was aware of the problems caused by inadequacy of supply. Government measures to control prices had not generally been successful and he blamed some cement manufacturers,

> "This is partly attributable to some manufacturers who, due to acts of omission or commission, did not apply proper penalty to those distributors who flouted sales directives."

He acknowledged that the final solution to rising prices was adequacy of supply and he hoped that the opening of Shagamu would contribute considerably to this solution.

He gave a final warning,

> "You are aware that cement has now been de-controlled by the Government and is now placed under the Resale Price Maintenance Agreement. The onus of ensuring that sellers and distributors comply with the agreed selling price will henceforth rest with the manufacturers. I hope, in fixing the selling price, the manufacturers will give due consideration to the convenience of the consumers. Our objectives will not be achieved and the economy would to that end be undermined if the producers and distributors enter into collusion at the expense of the consumers. It is a pitiful commentary on the attitude of the average businessman in Nigeria that they have, purely out of greed, astronomically increased the prices of certain items recently de-controlled. These people are plundering the economy and the people. They are killing them. Government objective is to encourage expansion of local production of decontrolled items not to put money in the pockets of privileged people who neither toil nor spin!"

The commissioning of the Shagamu works signified important developments at the Headquarters of WAPCO itself. As mentioned above, WAPCO started with a very modest headquarters establishment located first within the UAC Headquarters, and then in the Barclays Building at Marina. This created problems of ready access to the factory at Ewekoro and the division of administration and support services between the factory and the Headquarters. For a long time as much as possible was left at Ewekoro for that was "where the action was". The worsening traffic congestion of Lagos necessitated moving the Headquarters to Ikorodu Road, at Ilupeju, near the Oworonsoki Express Road out of Lagos, in 1975. It was then possible to separate the main task of manufacturing and quality control at Ewekoro from central Headquarters services. By the time of the commissioning of Shagamu, the Headquarters development with separate finance, technical, personnel, marketing and public relations departments had reached maturity. The Headquarters became the nerve centre controlling the action at the two manufacturing plants. Thus WAPCO made a quantum leap in terms of scale of production and evolved the administrative machinery able to move the Company forward to cope with the indigenisation programme then underway in Nigeria.

7

THE INDIGENISATION OF WAPCO

The Indigenisation Decree of 1972

Following independence in 1960 Nigeria moved gradually towards ensuring that the positions of authority and control in business, as well as in government, were occupied by Nigerians. In 1972 the government introduced The Nigerian Enterprises Promotion Decree. Its specific policy objectives and that of the indigenisation programme were to,

- create opportunities for Nigerian indigenous businessmen
- maximise the retention of profits
- raise the level of intermediate capital goods production

Chief Olu Akinkugbe, who, as we have seen, was soon to be on the WAPCO board and become its chairman, told a symposium organised by the Nigerian Economic Society to discuss the indigenisation decree in 1974,

> "I offer the following statement as incontestable: Nigeria needs a strong diversified growth economy, an economy in which Nigerians enjoy a rising standard of living, an economy which is industrialised and which at the same time can produce and distribute enough food to meet its needs, an economy which sustains its growth through the initiative of its people in mobilising its natural and capital resources."

As Chief Akin Ogunleye, who became managing director of WAPCO in the eighties, pointed out in his *Critical Analysis of the Indigenisation Decree and its Results* written for the National Institute for Policy and

Tony Jackson, John Milne, chief executive of Blue Circle, and Akin Ogunleye, later managing director of WAPCO

Strategic Studies in 1979, it was essential that the Nigerian economy was capable of relying on the initiative and resourcefulness of her people for self-sustained growth. It was also essential to promote a strong and diversified industrial economy.

To ensure the compliance of the provisions of the Indigenisation Decree, a number of institutions were established. The first was the Nigerian Enterprises Promotion Board and its related agencies. One of the agencies was the Inspectorate Unit whose officers were entitled to,

- enter business premises being used for any purpose to which the Decree relates
- inspect such buildings and books of accounts and other documents in order to ensure that the provisions of the Decree are being complied with
- obtain such information as may be deemed necessary for their purpose

Another was the Valuation Unit whose function was to determine and advise Nigerian businessmen on the value of the assets and liabilities of the enterprises they proposed to buy from aliens under the provisions of the decree. This was in order for the board to,

- take over, sell or otherwise dispose of such an enterprise
- distribute the proceeds of such sales or disposals

Another agency was the Bank for Commerce and Industry. The Federal Government had acquired 40% of the major foreign banks and insisted that they assist Nigerians to increase their interest in existing businesses, assist them to establish new business through promotion or acquisition and provide loans to buy from aliens selling their businesses. To help these endeavours the Government felt it was necessary to establish a separate bank, and the Bank for Commerce and Industry started operating on 2 October 1973, with an authorized share capital of N50 million.

The 1972 decree essentially divided the Nigerian economy into three parts. One part was reserved entirely for Nigerians, one part was set aside for Nigerian minority equity participation, and one part was not affected at all. All enterprises in businesses assigned to Schedule I of the decree were required to sell 100 per cent of their equity to Nigerian subscribers. No new foreign enterprises could be established in those areas after February 1972. Schedule II of the decree listed economic activities in which expatriates would be barred under certain conditions. Small firms, defined as those with equity under 200,000 Nigerian pounds ($608,000) or sales turnover under 500,000 Nigerian pounds ($1,520,000), were required to sell all of their equity to Nigerian subscribers. Large firms, which accounted for nearly all of the enterprises operating in these sectors, were required to make 40 per cent of their equity available to Nigerians. The rest of the economy, which included WAPCO, was not affected by the decree at all.

Owing to the size of its equity shares and annual turnover, WAPCO belonged to the range of businesses not affected by the Decree. However, while the Decree reflected the anxiety to promote the participation of Nigerian businessmen in equity shareholding, Nigerian staff in the supervisory and management cadres increased their pressure for the indigenisation of various positions and equalisation of conditions of service between expatriate and Nigerian staff. The management of WAPCO under Jackson had begun to respond to such pressures even before it became a feature of the second Indigenisation Decree.

Was the 1972 Decree successfully implemented in practice? Chief Ogunleye showed in his Analysis that by the end of 1974 the transfer of small businesses and the services sector – those reserved exclusively for Nigerians – had taken place in 237 out of 326 businesses with a value of N20 million. Out of 628 businesses where aliens were barred in certain circumstances 503 worth N60 million had complied.

It had been calculated that an extra 1,000 senior managers and 2,500 managers would be required each year to cope with the changes. In the event, these calculations proved an underestimate because of the growth in the economy, and the Government established the Administrative Staff College of Nigeria, the Centre for Management Development, the Industrial Training Fund and many other management training facilities to meet the manpower needs.

In spite of these successes Chief Ogunleye concluded that,

> "The Decree had achieved the objective of raising Nigerians' capability in the development of intermediate technology. It was, however, far from being satisfactory and Government, therefore, still had to look for ways and means of improving the opportunities for generating the effectiveness of her policies regarding indigenisation."

The next stage of Indigenisation

By 1974 the Gowon regime was becoming unpopular and the reneging of his promise to return the country to civilian rule by 1976 was for many the final straw. Reports of corruption were rife, with the local military governors seeing their territories as personal chiefdoms and treating them as such. For many of his colleagues in the army, however, it was the indecisiveness and inefficiency, symbolised by the traffic chaos in Lagos and the cement armada, that were the most intolerable. In July 1975 Gowon was overthrown in a bloodless coup while he was in Kampala at a conference of the Organisation of African Unity (OAU) and was replaced by General Murtala Muhammed. After the failure of an attempted coup in February 1976, in which he was assassinated, he was replaced by his deputy, General Olusegun Obasanjo. Among the main supporters of the regime were a young and radical group anxious to impose public sector control on a private sector they felt had not been amenable enough to Nigerian national control during the post Independence decade and the oil boom years. Their feelings were summed up by Joe Makoju, currently assistant managing director, who said,

> "For any nation to survive its people must strive to direct their own lives as much as possible and to reduce their subjugation to the power and interests of other societies outside the demarcation of their nation."

General Murtala Muhammed who replaced General Gowon and tried to push for indigenisation of the Nigerian economy

The new Government felt that the State governments had grown too powerful and proceeded to use the increased earnings of the Federal Government from the oil boom, and the centralised control natural to a military regime to assert Federal Government control over the states. Some of the major developments of the Western Region in the sixties were taken over by the new Government, for example, Western Nigerian Television – the first television station in Africa – and the University of Ife. In an attempt to reduce their power further, the Government increased the number of states from 12 to 19.

The Western State was split into three: Oyo, Ondo and Ogun states. The former investments of the Western State, including their shareholding in WAPCO, were placed in an investment holding company, the Odu'a (short for Oduduwa, the mythical progenitor of the Yoruba) Investment Company Limited incorporated in July 1976. Besides an interest in WAPCO this investment company also inherited interests, some of them controlling, in the National Bank of Nigeria, Lagos Airport Hotel, Great Nigeria Insurance Company,

Cocoa Industries, Vegetable Oils (Nigeria), Nipol, Askar Paints, Wemabod Estates, Nigerite and Dunlop Nigerian Industries. The general aims of the Odu'a Investment Company were to,

> "Engage in activities which will promote and enhance rapid economic industrial and human resources development in Nigeria.
>
> "Pursue economic and social goals as defined by the Board.
>
> "Assure better sector and enterprise arrangements by means of incorporation of and partnership in new ventures, merger dissolution or take-over of existing companies."

The board of the holding company was constructed in a manner that was designed to be equitable. The board was made up of representatives of the three states with the chairmanship rotating amongst the three states.

The result of the analysis of the situation by the new regime was a second Indigenisation Decree in 1977 designed to improve on the 1972 decree in scope and implementation. A Price Control Board was introduced as well as a Productivity and Incomes Board which monitored the progress of indigenisation, expatriate quotas and the prices of major products in the private sector. All businesses involving aliens were affected. They were now classified into three categories and indigenisation was either total or partial. The number of types of business in the wholly owned category – schedule 1 – was raised from 22 to 35. Schedule 2 gave a list of businesses in which Nigerians must own 60% of the equity shares and Schedule 3 contained the businesses in which Nigerians must control 40% interest.

> "The coverage of the 1977 Decree was, therefore, total. The intention could not but be clear. In short there could be no economic activity going on in the country in which Nigerians have no stake. In terms of technology, financing, management capability and personnel, investment opportunities and activities must have built into them the extent to which Nigerians would contribute to the successful execution of projects."

To implement the Decree more effectively the Capital Issues Commission was to ensure that,

- Nigerians know what they are buying
- Nigerians pay reasonable prices for what they buy
- the operations of those businesses purchased are sustained commercially

The Government increased its stake in alien banks to 60% and gave them guidelines to ensure they liberalised their lending policies to give Nigerians access to funds essential for the acquisition of equity stakes in alien firms.

In Ogunleye's view this last aspect of the Decree, which was designed to increase opportunities for indigenous participation, in fact increased the inequality of opportunities. Only the few rich had the necessary collateral to gain access to the banks to obtain loans for the acquisition of shares. The only mitigating factor was the clause in the Decree which reserved ten per cent of the shares of enterprises in schedules 2 and 3 for the staff. And he was not only critical of this financial aspect, but also of the "capability to control" aspect,

> "It is my view and experience that indigenous technical know-how is absent as most of the multi-nationals have their Research and Development set-up centralised in their home countries."

On the training of Nigerian personnel Ogunleye thought it essential that,

> "There should be concerted effort to ensure that where expatriates are required to provide technical services for which Nigerians are not available, programmes for training Nigerians are agreed upon thereby ensuring the desired transfer of technology and technical know-how. It is only by doing this that we can gradually put Nigerians in positions of authority with sufficient responsibility to make them assume the 'control of the commanding heights of the economy.'"

In his *Critical Analysis* Ogunleye concluded,

> "The Indigenisation Decrees have indicated directions of policy and shown guidelines to enable Nigerians to aspire to take advantage of the opportunities available to exploit our resources. It is my view that they are not sufficiently articulate and comprehensive in terms of the philosophy of economic development such as to make us control the commanding heights of the economy now or

in the near future. They are, however, a good beginning on which future policy could be developed."

WAPCO's reaction to Indigenisation

As we have already seen, one of Nigeria's leading businessmen, Chief Olu Akinkugbe, was full of praise for WAPCO's attitude to the aspirations of Nigerians to raise their skills – and therefore standard of living – and to control their own affairs. From the earliest days the company had been training personnel at all levels and in 1975 appointed a Nigerian chairman. In the same year David Adeyemo was appointed as the first Nigerian general works manager at Ewekoro, an appointment which the Management Staff Association regarded as a major breakthrough in the struggle for Nigerianisation. Adeyemo had been to Government College and the University College, Ibadan, before going to Manchester University in the UK where he graduated with an honours degree in chemical engineering in 1957. He spent three years with APCM in London before

Nze Oliver Nwaozuzu – promoted to finance director of WAPCO in 1975

returning to Nigeria in 1960 to become chief chemist at Ewekoro. He was promoted to production manager and then to deputy general works manager before his appointment as general works manager.

Also in 1975 Oliver Nwaozuzu was appointed finance director. A chartered accountant and associate member of the Association of Certified Accountants, the Institute of Chartered Secretaries and the Institute of Bankers, Nwaozuzu graduated in economic planning and was working in the UK as chief group accountant at Claude Rye Ltd when he was recruited for WAPCO by Tony Jackson. He returned to Nigeria to join WAPCO as financial controller in 1974.

By the end of the seventies the total WAPCO staff strength had increased to 2,500. There were 160 Nigerians in managerial positions compared with only 17 in the sixties, 43 in 1972 and only 88 as late as 1975. By 1980 there were 40 expatriate managerial staff performing specialised and technical services and on short term appointments By 1990 this was reduced to approximately six per works principally engaged in specialised support roles only.

The public flotation of WAPCO

As part of the indigenisation process WAPCO floated itself as a public company on the Nigerian Stock Exchange in 1978. The Nigerian Stock Exchange had begun life in 1961 as the Lagos Stock Exchange. It followed the example of the London Stock Exchange in setting up systems although it did not have the jobber/broker system but rather the single capacity method of operation eventually introduced in London in Big Bang in 1986. The Federal Government had become involved as a shareholder in financing the building of the Shagamu factory and with the Western State held 51.7% of the equity. To comply with the Nigerian Enterprises Promotion Decree of 1977 the offer to the public of shares would take the total Nigerian stake to the required 60%. The individual shareholdings became BCI 33.4%, UAC 6.6%, Odu'a 26.8%, Federal Government 16.6% and the public 16.6%.

The construction at Shagamu had cost N106 million (at N1.16 to the pound that was £91 million, or in 1990 terms about £300 million). The layout of the factory had deliberately allowed for further expansion, and the demand for cement and the desire to reduce the outflow of foreign exchange were such that the Federal Military Government was already encouraging the company to install a third kiln at Shagamu. As the offer prospectus put it,

> "This is a natural development for the company and the

The WAPCO offer of shares to the public

> design work is at a very advanced stage. The negotiations with machinery suppliers have been virtually concluded and overseas credit has been secured on terms which are considered to be satisfactory and in conformity with Government policy on funding the expansion scheduled to begin towards the end of 1978, which is estimated to cost N52 million ... Management anticipates that on completion of the proposed expansion, total production capacity of the company's factories at Shagamu and Ewekoro should be about 1,800,000 tonnes per annum."

The company expected to make N9.5 million before tax in the year ending 30 November 1978 and proposed to pay a dividend of 11.7% on the enlarged capital of N30.15 million. At the offer price of 55K per share this would give a yield of approximately 10.6%. The prospective earnings yield would be 26.7% representing a price earnings ratio of 3.75 times. (For comparison, Blue Circle's ordinary shares at 231p on the London Stock Exchange yield 5.8% and have a price earnings ratio of 9.6.)

The day of flotation

As well as a Nigerian chairman, Olu Akinkugbe, the board consisted of six other Nigerians of whom two were executive compared with five British directors of which two were executive. The company was well on the way to being indigenised and ready for further expansion.

8

WAPCO IN THE EARLY EIGHTIES

Civilian government returned to Nigeria in 1979 after 13 years of military rule and revived the intense rivalries of the past. The Unity Party of Nigeria (UPN), the successor of the Action Group of the sixties, won elections in all three Yoruba states and was anxious to continue the development and welfare programmes of the Action Group. It was also anxious to assert the powers of the state governments against those of the Federal Government – again an alliance of the National Party of Nigeria (NPN) and the Nigerian

Tony Jackson showing the Akarigbo and some of his chiefs round the Shagamu site. It was from him Jackson received his chieftaincy

G H Wilson, the UAC representative on the WAPCO board, with Gus Townshend, commercial director

Peoples Party (NPP), the successors to the NPC and the NCNC of the 1960s. The Federal Government on the other hand wanted to continue the centralisation policies of the military regimes at the expense of state governments.

Immediate effects of the opening of Shagamu

How did the efforts of those who were toiling and spinning at the new Shagamu factory affect WAPCO's performance? Naturally there was an increase in turnover, but 1979 was a difficult year. The country as a whole experienced a recession, reducing demand. The Comprehensive Import Supervision Scheme with the notorious Form M may have reduced cement imports, and this was of benefit to WAPCO, but it also held up vital spare parts which was definitely not of benefit. Furthermore the rainy season persisted much longer than usual making quarrying difficult. Compounded with the shortage of equipment, output from the quarries and therefore the factories was curtailed. Costs also increased as the company was forced to bring in outside contractors to help shift the overburden. As a result, although turnover increased by 44%, profits did not rise in the same proportion. In his annual report to shareholders, the new chairman, S O Ige, complained that the company's return on investment at 6.6% was insufficient to support the company's long-term plans for re-investment in new machinery. He complained of

Tony Jackson, David Stirling from Blue Circle, with Chief and Mrs Akintola Williams. Akintola Williams was one of the company's external auditors until he retired in 1988

the refusal of the Price Control Board to allow the company to pass on the increased costs. The first price rise for three years had been allowed in April 1978, but by June 1979 a further increase had been necessary and the application made to the Board. By May 1980 in spite of constant pressure this increase had not been granted. Ige said in his statement,

> "Last year we produced 50% of the cement made in Nigeria and therefore our present contribution to import substitution cannot be in doubt. While we appreciate the need for Government to curb inflation, we must be allowed to regulate our prices if we are to continue to fulfil our role in the economy of the country."

He was pleased to note that consultations with Government had led to a modification of the Comprehensive Import Supervision Scheme so that industrial raw material (in WAPCO's case, gypsum) and spare parts were excluded from pre-shipment inspection. The company had also been in consultation with the Government over the setting up of more depots to enable them to ship directly to consumers and users. This was a response to the feeling that many of the distributors were using the tight supply situation to push up prices, thereby fuelling inflation. WAPCO planned to open more depots in its

S O Ige, who trained as a barrister, was appointed chairman of WAPCO

existing market areas and new ones in fresh markets in the north-western part of the country.

A new chairman

Ige himself came to WAPCO from his position as a member of the House of Representatives and was appointed to the WAPCO board to represent Odu'a's interests and was made chairman following the resignation of Akinkugbe. He read law at the London School of Economics from 1953 until 1956 and was called to the bar in 1957. He returned to Nigeria in 1957 and worked for a year for Shell as one of their first group of Nigerian graduates. He then set up in private practice as a lawyer and attended post-graduate courses at Harvard University and the University of Chicago. He maintained his interest in business and held a number of directorships during the sixties and seventies including the chairmanship of Leventis Technical Ltd. He was also involved in politics, first as a councillor in local government, secondly as a legal adviser to the British High

Commission and finally as a member of the House of Representatives through the UPN.

His first full year as chairman of WAPCO coincided with one of the very few periods in the company's 30 year history when the demand/supply situation was almost in balance, and Ige was forced to comment in his 1980 annual report to shareholders,

> "We were not able to increase our sales price to the profitable level we would have wished, due not only to Government restraining posture *but to natural prevailing market constraints imposed particularly by cement imports."*

Turnover barely increased from N74.7 million to N75.4 million and profits also remained static. A company as large as WAPCO and as deeply involved in production tied to a central economic activity of the nation, construction, was bound to be affected by the prevailing economic conditions and government decrees on taxes, pricing policies, credit restriction and imports. A further sharp rise in the price of oil led to hopes of increased economic activity in Nigeria. This did not materialise and WAPCO's sales and profits were restrained by their inability to raise prices to match cost increases, exacerbated by the introduction of a statutory minimum monthly wage of N100 as well as general wage increases of 10-15%.

WAPCO celebrates 21 years

On 5 December 1981 WAPCO held a celebration to mark 21 years of production of cement at Ewekoro. Dr I J Igbani, the Minister of State, Federal Ministry of Industries, was invited and took the opportunity to spell out the government's attitude to industry in general and cement in particular. He praised WAPCO for its 21 years of effort, contribution to the Nigeria economy, use of local resources, human and material and its transfer of technology. He emphasised how critical cement was in the development of the country and said that the government planned to make Nigeria self-sufficient in cement within the foreseeable future. Government investment in three new cement plants was planned as well as encouragement to existing manufacturers. He also talked of protection,

> "The Federal Government has taken note of the complaints of the Cement Manufacturers Association of Nigeria that the Government's open door policy aimed at removing scarcity may be hurting the local producers. The Government has, therefore, decided to set up a

> Cement Monitoring Unit in the Federal Ministry of Industries, of which your Association is a member. The unit will monitor activities in the cement sub-sector including inflow, distribution and local production, and advise the government from time to time on the volume of cement import as well as on the local production of cement. This is to ensure that the industry does not suffer as a result of the government's desire to make cement available at a reasonable price for the housing and construction projects. Cement imports will be permitted only in order to complement local supply and prevent the country from being turned into a dumping ground for all sorts of foreign cement."

The Minister pointed out that the government was doing everything possible to assist local producers. For example, it had recently directed that cement paper bags and selected spare parts could be imported free of duty. The government was also looking into the possibility of establishing a plant to manufacture gypsum locally.

Ige replied to the minister and congratulated the Federal Government on the success of its co-operation in this venture with the state government, private industry and the private investor. He went on to say that,

> "The working relationship amongst the principal shareholders has been cordial. Ours has been an experiment in industrial co-operation namely between the state governments owning Odu'a and the Federal Government and the result has been commendable. I would therefore urge that the Federal Government should as a matter of deliberate policy involve itself in more industrial undertakings with state governments. Apart from the obvious advantages this can bring it will help a great deal in promoting understanding and harmony between it and the state governments."

Certainly the rivalry between Federal and State governments had not been helpful to some industrial concerns in Nigeria. But no such difficulties arose within WAPCO which continued to work with both and continued to build a successful and profitable cement factory.

A new appointment – vice chairman and chief executive

Tony Jackson, by now Chief, The Muleoruwa of Ijebu-Remo,

Chris Akande (at microphone) being sworn in as Western State Head of Service in 1975. He was to become chief executive of WAPCO. With him is David Jemibewon, Military Governor of Oyo State

Jackson, had returned to the UK after five years in Nigeria during which time he had taken WAPCO steadily along the path of indigenisation and had master-minded the public flotation of shares. (He was awarded an OBE for his services to the business community and community relations in Nigeria.) Before he left, arrangements were put in hand following consultations with Chief Akinkugbe to invite Chris Akande as executive vice chairman with a view to maintaining the momentum of indigenisation and to try and ensure that, with the return to civilian rule imminent, the executive leadership was in the hands of an apolitical, experienced Nigerian. The expatriate technical director, Ron Burnett, was appointed managing director. His early career had been in mining and in the heavy end of the electrical industry as an engineer. In the early seventies he had been deputy southern area production manager at APCM and then PA to Gordon Marshall who was at that time works director. He had been appointed technical director of WAPCO in 1976.

Akande was an engineer by profession. He had been one of the first undergraduates at Ibadan University when it was founded in 1948 and from there won a scholarship to the UK. For a year he was an engineering assistant at the Essex Rivers Catchment Board at Chelmsford and in 1950 he went to Battersea Polytechnic for three years. In 1953 he joined the consulting engineers, Ove Arup and Partners, returning to Nigeria in 1955 where he was appointed to the Western Region Public Works Department as an executive engineer. In 1966 he became chief civil engineer and in 1967 controller of works and the principal technical adviser to the Western Region Government. During the civil war he enjoyed plenty of excitement, including the blowing up of bridges to halt the invasion of the Biafran armies. After the war he returned to the Ministry of Works and in 1973 became Permanent Secretary. After the 1975 coup he became Secretary to the Government and head of the Civil Service in the Western State. With the creation of the three states, Oyo, Ondo and Ogun in the Western State Akande was to have been appointed head of the civil service in his home state in Ondo, but instead was requested by the Head of State to stay on in Oyo. Soon after that he left the civil service and agreed to join the Odu'a Investment Company, which had taken over the role of the Western Nigeria Development Corporation (WNDC). Although he was offered an executive director post in WAPCO in 1976 it was not until 1979 that he resigned as managing director of Odu'a and joined the company, becoming vice chairman and chief executive in October.

Akande was, and is, a straight-talking, upright individual who gives his all to whatever he sees as the task of the moment. As the first Nigerian chief executive he saw his task as studying WAPCO's overall operations in detail with a view to re-organising them while leaving Ron Burnett to run the company day to day.

> "I joined the company with a great sense of mission. While my appointment as the first Nigerian chief executive was in furtherance of Nigerianisation of posts in the company, I saw the need to use my previous experiences to chart a new management style for the company. I also saw the need to have laid down proper procedures to ensure uniformity of standards in all facets of the company's corporate management with a view to establishing, as far as is possible, high efficiency and high standards of probity. This meant reorganising the company's management structure at all levels to provide a meaningful succession plan including the Nigerianisation of posts at all levels."

Akande flew to study Blue Circle in the UK and went out to see their operation in Malaysia. He returned and, after further study of the company in Nigeria, put forward a Memorandum of Re-organisation. It recommended that each division have a clearly defined function

Ron Burnett – technical and then managing director in the late seventies and early eighties

and that a management *team* should be created so that all heads of department would understand what was happening throughout the company. This was a transition period and the top management structure would need adjustment to avoid overlap and consequent misunderstandings. Tolerance was required.

When Burnett returned to the UK and was replaced by the expatriate managing director of Ashaka cement in the north, Colin Roots, matters began to come to a head. It was at this time that Tony Jackson was appointed joint chief executive of Blue Circle Overseas and became involved again in WAPCO's affairs. Roots returned to the UK and Akande assumed full control as managing director and chief executive, supported by Dr Ransom Murray, who was brought out from his position as general manager of Blue Circle's research division to the new post of deputy managing director. It has been one of WAPCO's outstanding appointments in that Murray achieved a remarkably close relationship with his colleagues and in particular with the managing and finance directors. For five years the three directors have successfully brought the company through many financial and technical difficulties and provided a model of leadership, integrity and effective communication, which many other industries might wish to emulate. Dr Murray brought with him

Sir Rowland Wright, chairman of Blue Circle, with Chief Ernest Shonekan, chairman of UAC (Nigeria) in 1982

(From left to right) Chris Akande, Dele Ige, Colin Roots (managing director for 18 months in the early eighties) and Oliver Nwaozuzu

not only many years' scientific and technical experience of the cement industry but also previous experience of conditions in West Africa assimilated during his period as assistant director, Ghana Geological Survey. He had been recruited in 1953 by the same Dr Dixey of the Directorate of the Colonial Geological Surveys mentioned in chapter two.

Dr Ransom Murray, deputy managing director

Dr Murray's wife, Dorothy, who carried on the tradition of charitable work

Jackson, a former chairman and managing director, was invited by the WAPCO board to become non-executive vice chairman and shortly afterwards Akin Ogunleye was appointed assistant managing director with a view to succeeding Akande. Unrelated to these management changes Odu'a determined to change its board representatives and Ige himself was forced to resign to face charges from the new military regime, arising from his involvement in politics.

Difficult times

In the meantime how had the business of making and selling cement been progressing? At the end of 1980 the unprecedented happened – the factories were closed for three weeks owing to lack of demand. At the end of the first six months of the year to November 1981 profits were N1 million against a budgeted N6 million. In the second half of the year demand improved and the company was ultimately frustrated to find that it could not meet demand owing to power outages and plant breakdowns. The company also suffered from industrial disputes both internally and in the country at large. In the event, turnover increased by 26% to N95 million, while profits increased only marginally.

By the early eighties the Nigerian economy was severely out of balance. Dependent by then on one commodity, oil, when the price of that commodity fell, as it did from 1981 onwards, the country's trade balance lurched deep into deficit. Furthermore the government had borrowed a great deal from overseas countries in the heady days of rising oil revenues. By 1982 the *Financial Times* was writing, under the headline,

> "Nigeria – Africa's biggest market under strain,
>
> "The Government of President Shehu Shagari is engaged in the sort of delicate balancing act which will undoubtedly tax its ability. It is seeking to maintain the confidence of the international trading and banking community while depriving it of timely payment for foreign goods ... It wants local industry to keep going but is cutting back its supply of imported raw materials and hitting its cash flow by compulsory import deposits. It is attempting to control the economy with a bureaucracy which, at best, is slow to move and, at worst, is corrupt to the point of undermining Government policy for financial gain."

One of the government's remedies was to pay foreign suppliers only what it received each month. This was all right for the present but it meant a host of unmet claims from the past.

In early 1982 the pressure exerted on the Federal Government by the Cement Manufacturers Association of Nigeria over cement imports was rewarded by the imposition of a 20% import duty, and WAPCO was able to contain its own costs by the reduction of 264 permanent and 365 casual staff. Unfortunately, importation of spare parts was placed under licence. Parts needed for the plant, and more importantly for the generators which WAPCO were busily trying to install to combat the continuing failure of NEPA to maintain constant supply, were held up. The situation was made worse by the difficulties in getting Letters of Credit confirmed under the Economic Stabilisation (Temporary Provisions) Act so that the two 3.5 MWA electricity generators could not be installed as planned in October 1982. Indeed Ige was telling shareholders in June 1983 that it was not certain they would be ready by August 1983 as a vital Letter of Credit had been confirmed only in March.

The import restrictions also affected the supply of cement paper bags as the local manufacturers struggled to import the necessary Kraft paper. WAPCO imported five million plastic sacks, but still lost the sale of 53,000 tonnes between July and September 1982.

In his annual report to shareholders in the middle of 1983 chairman Ige sounded quite desperate,

> "The Cement Manufacturers Association of Nigeria put in a concerted representation to all relevant areas of Government for assistance. Although once again the Government through the Special Committee listened to our plea, very slow administrative process did not make it possible for us to receive promptly the much needed import licence. It is only hoped that we will not be forced to shut down our operation in due course. We plead with Government to help us to help the nation. If local cement manufacturers grind to a stop, it will mean that all the country's cement needs will have to be imported thus losing substantial foreign exchange."

He went on to plead for the speedy issuing of the necessary import licences.

Progress in the eighties

By the middle of the eighties WAPCO was still responsible for about 50% of the country's indigenous production of cement. Other plants, with the exception of Ashaka, were not coping well with the difficult circumstances of worldwide economic recession and Nigerian import constraints. Projected demand for cement was falling like a stone. In 1983 the industrial consultancy unit of the Nigerian Institute of Social and Economic Research had projected a demand for 1984 of 8.9 million tonnes, assuming a five per cent growth rate. In fact, demand was only five million tonnes. Even so there was no relief to consumers because supplies only reached 75% of the lower demand. Chris Akande estimated that only three million tonnes would be produced by the seven manufacturing plants in Nigeria. This was the same as 1983 despite higher installed capacity. Furthermore, imports plummeted from 3.7 million tonnes to 1.3 million.

Of the three million tonnes produced the two WAPCO works again produced half. Even WAPCO continued to suffer from the shortage of foreign exchange and was unable to purchase the spare parts necessary to keep the two plants fully operational. Just after the Annual General Meeting in 1984 the board appointed Otunba Oduntan as its new chairman and he was able to report satisfactory results for the calendar year 1984 (WAPCO had now changed its year end to 31 December). Pre-tax profits had increased from N21 million to N28.99 million. Shagamu was going full blast and exceeded its installed capacity to produce just over one million

Otunba Victor Oduntan – followed Ige as chairman

tonnes of finished cement – an outstanding achievement. This had been made possible by the recognition that the guaranteed output from the kilns could be exceeded substantially if they could be fed with sufficient raw materials. It was therefore decided to install a third raw mill. This enabled clinker output to rise from about 650,000 tonnes in 1981 to 800,000 tonnes in 1982 and 925,000 by 1988. Ewekoro, now nearly 25 years old, suffered heavily from the shortage of spare parts and produced only 500,000 tonnes against an installed capacity of 750,000 tonnes.

Otunba Victor Oduntan was born in 1934, a royal descendant of Oba Tunwase who reigned in Ijebuland from 1886 to 1894. After five years at Ibadan Grammar School he joined the Western Nigeria public service, from where he won a scholarship to the University of Exeter and graduated in economics and government. He then won a master's degree in development economics at Williams College, Massachussetts. He returned to Nigeria to work in various capacities in the Western Nigeria public service and became Permanent Secretary in 1975. When Ogun State was formed he served as the first Permanent Secretary, Ministry of Finance and Economic Development. In 1979 he was appointed Secretary to the Military Government and Head of Service from which post he retired in 1984. During his career in the civil service Oduntan had served on the board of many companies, including the National Bank of Nigeria, Guinness Nigeria, Tower Aluminium Nigeria, CSS Bookshops, Western Nigeria Marketing Board, Nigerian Petroleum Refining Company and Odu'a Investments Company. He had also written some articles in leading journals on economic planning and been a major contributor to the Second and Third National Development Plans.

In early 1985 the company was laying down plans for an expansion of Ewekoro to bring the installed capacity of the two factories to over two million tonnes. It was estimated that a suspension pre-heater dry process plant to produce 600,000 tonnes of clinker per annum would cost N100 million and the company was looking at the government to make available the necessary foreign exchange. It calculated that every naira of foreign exchange spent on spare parts for the indigenous cement manufacturer saved six naira on imported cement. WAPCO had already set up a division, Portland Electrical Repairs, which was saving foreign exchange by carrying out repairs of electrical equipment in Nigeria, and, as a member of the Cement Manufacturers Association of Nigeria, it was exploring to see if gypsum could be found in the country in commercially exploitable quantities.

Nigeria itself was struggling with heavy overseas debts incurred in the boom years of the seventies. The early eighties were anything but boom years and the price of Nigeria's one precious commodity, oil, had fallen dramatically since the heady days of 1979. Oduntan was suggesting a new remedy in his 1984 annual report to shareholders,

> "We support the Federal Military Government's policy of economic belt tightening through austerity measures to curb the catastrophic increase in debt load as well as to provide funds to service existing debt. The country must appreciate the need for continued self-sacrifice and increased productivity if we are to succeed in revamping the economy. An overvalued currency cannot be in our country's long-term interest. A studied, guided and well monitored devaluation of the naira might be inevitable in the long run."

However, there was to be no help for the beleaguered Nigerian economy from events in the outside world. The price of oil continued to fall throughout 1985 and in early 1986 fell to a level not seen since the early 1970s. This made it very difficult for the government to meet its debt service obligations and provide sufficient foreign exchange for even the raw materials essential for the full utilisation of installed productive capacity. The government was forced to resort to counter-trade licences, effectively barter. In a relatively undeveloped economy such as Nigeria's most of these licences could not be used and the results for home-based manufacturers, including WAPCO, were little short of disastrous.

Nevertheless, thanks to the superhuman efforts of its management and employees, WAPCO managed to maintain its production close to capacity levels. This involved a special licence granted personally by the President after a case was made by Chief Ogunleye. He told the President the company would have to stop production unless essential spare parts were imported. The production staff at Shagamu had to improvise to keep the raw mills running by cutting up disused railway lines and reinforcing steel rods into small pieces due to the shortage of grinding media. When the licence was granted a Boeing 707 was chartered to bring in 40 tonnes from Italy as a stop-gap while the remaining 400 tonnes was brought in by sea. Profits crept ahead to N30 million.

During 1985 two major events occurred within WAPCO. It celebrated the Silver Jubilee of the company's formation, and it appointed Akin Ogunleye as its chief executive. Ogunleye had graduated in economics at Southampton University in the late fifties and had returned to Nigeria in 1961 where he began work as an administrative officer in the Western Region civil service. He continued as assistant secretary in the Ministry of Education and later as private secretary to the Governor. He also worked in different capacities in the Ministry of Trade and Industry, the Ministry of Economic Planning and Social Development, in the Premier's office and at the Ministry of Finance. Later he worked as a Division officer, Ekiti Central Division, and at the Ministry of Establishment and Training, at the Ministry of Health and finally as Under Secretary at the Ministry of Economic Planning and Reconstruction.

During this wide experience in the civil service Ogunleye found time to take postgraduate courses at the Universities of Pittsburgh in the USA and Ibadan. He attended the inaugural course at the National Institute for Policy and Strategic Studies (the President, then Brigadier Ibrahim Babangida, was on the same course), the Henley Management College and the London Business School. He also became a member of the Nigerian Institute of Management and took its top management course in 1979. He joined WAPCO thanks to a chance conversation in 1973 with Ron Horn, then managing director of Cocoa Industries Ltd, in which he said he was considering leaving the civil service and joining private industry. This information was passed on to Tony Jackson, who had just begun his search for able Nigerians to replace the expatriate managers in WAPCO. Ogunleye's wide experience in civil service jobs, mostly involved in economic expansion, made him an ideal choice.

Jackson could see that Ogunleye had thought through the ramifications of Nigeria turning itself from a backward agrarian society into a modern industrial nation. Ogunleye had been totally in accord with the aims of the Western Region in setting up industrial estates – Ikeja on the outskirts of Lagos was the most notable – in order to create employment, develop technical know-how and transfer technology. There was some debate whether the businesses developed should be large or small, but the ultimate decision was that the most important criterion was import substitution. Hence the Western Region became financially involved with WAPCO, Nigerian Textile Mills, Dunlop, Cocoa Industries and the palm kernel industry. The other important criterion was the local availability of raw materials.

Nigeria was anxious to increase her technological know-how and wanted technical agreements. WAPCO's agreement with Blue Circle proved to be immensely satisfactory and worked particularly well in the training of Nigerians. Ogunleye himself followed his general experience in the civil service with a general experience of WAPCO. He started in February 1974 as the manager of administration at Ewekoro works and in 1976 went to the UK on a stores management course. He returned as procurement manager and in 1979 became assistant commercial manager to the commercial director, Gus Townshend. At this time he was nominated to go on a course at the Manchester Business School but in fact went on a senior executive course at the National Institute for Policy and Strategic Studies in Kuru, near Jos. (We have already seen extracts from his Critical Analysis on the effects of indigenisation.) As well as being beneficial in itself this course enabled Ogunleye to meet some interesting and potentially influential people. For example, the current President of Nigeria was on the same course and in the same syndicate. When he returned to WAPCO he did not succeed Townshend (who had been seconded from UAC and served for over 30 years in Nigeria) but became sales and marketing director before becoming chief executive in 1985.

In 1986 Odu'a again withdrew some of its nominees including the

Chief Joseph Ale – appointed WAPCO chairman in 1986

chairman, Oduntan. He was replaced by Chief Joseph Ale, who had been a director of WAPCO since 1984. A career civil servant he had also served in the foreign civil service with appointments in Ethiopia, London and Paris. He retired from the civil service in 1981 and took up a business career in property. During Ale's period as chairman he faced considerable difficulty in balancing guidance from Odu'a with his duties as a director of WAPCO and particularly as its chairman. The company is greatly indebted to the manner in which he showed great integrity in these circumstances. Following his resignation the shareholders reviewed the way in which continuity of board representation might be improved particularly the position of chairmanship and more appropriate arrangements have been devised.

A new economic programme

By 1986 it was clear that only a radical alteration of government policy would stand a chance of bringing Nigeria out of its economic slump. The import licensing system had led to the decline of industrial output to between 20% and 30% of installed capacity. The foreign debt of the country had moved far beyond its capacity to pay.

Under these circumstances the import licence scheme was replaced by the Second-Tier Foreign Exchange Market, effectively a massive devaluation of the naira. This meant that imports became expensive but at least companies had some chance of finding the foreign exchange to bring them in provided they had enough naira to buy at the inflationary prices.

What it meant for WAPCO's plans to re-develop the Ewekoro works was a little more far-reaching. Plans had reached an advanced stage but the Second Tier Foreign Exchange Market had pushed the cost in naira up from N100 million to N1 billion. The management had to reassess the situation and a plan evolved of installing a third kiln at Shagamu while refurbishing Ewekoro so that it could at least maintain current levels of output, or perhaps slightly exceed them. The long-term plan was to generate sufficient funds to tackle the major Ewekoro redevelopments at a later date – it is with Ewekoro with its huge reserves of raw materials that WAPCO's long-term future lies.

9

THE STRUCTURAL ADJUSTMENT PROGRAMME

The Nigerian economy

By early 1986 the squeeze on the Nigerian economy imposed by the falling oil price was intense. One businessman said,

> "I have written this year off. I will just close my eyes and keep on going."

The *Financial Times* was equally pessimistic,

> "For any manufacturer in Nigeria today, the prospect is grim indeed. By common consent only the biggest can survive the shock that is to come ... The reason for this pessimism is of course Nigeria's fearful dependence on one commodity – oil. So much of the benefit of oil has been squandered, so much money has been siphoned off into external bank accounts that the Third Oil Shock – as the present price slide is being called – will be twice as painful as it should have been.
>
> "Oil has been allowed to squeeze out almost every other export industry. Last year it accounted for more than 97% of foreign exchange revenue ... Agriculture was allowed to become moribund. Nigeria used to be among the world's biggest exporter of groundnuts. Today it has to import them."

The sharp drop in the price of oil had created a balance of payment crisis. Nigeria was finding it difficult to service its debts and provide the foreign exchange necessary to pay for essential imports. The civilian government had tried to meet the situation by restricting

President IB Babangida – he introduced the Structural Adjustment Programme

imports through a process of licensing, but had only compounded the problem through massive corruption in the allocation of licences such that manufacturers who needed them for essential raw materials and spare parts did not obtain the import licences, whereas touts who only wished to sell to the highest bidder received licences. Much foreign exchange was also allocated for the importation of essential food such as wheat and rice, opening further room for corruption and political jobbery. The military regime under Major-General Mohammed Buhari, which took power at the end of 1983, concentrated on cleaning up the political process without significant changes in the economic policies of the government it had displaced from power. In particular, it refused to take up the alternative of progressive devaluation of the naira advocated by Chief Oduntan, the former chairman of WAPCO, and being increasingly demanded by Nigeria's trading partners as the minimal condition for rescheduling the massive debts. The government of General Ibrahim Babangida, who seized power in August 1985 with the title of President, could hardly postpone coming to grips with the problem.

The Structural Adjustment Programme

The problems were formidable – inefficient and misconceived projects were left over from the oil boom days, agriculture was backward, industry was operating at about 30% of capacity. Looming large over every other problem was Nigeria's exploding population. As the *Financial Times* put it,

> "The implications of the population growth rate of over 3 per cent represent one of the serious challenges faced by the country. At the present rate of increase, Nigeria's population will increase nearly three-fold by 2015 to more than 280 million, placing an enormous strain on the country's resources. ... No sector of the economy is left untouched by the consequences of the dramatic growth in the population, whether it is the growing proportion of oil which has to be reserved for domestic use rather than export, the ecological consequences of deforestation as a result of rural energy needs, or the inability of commerce and industry to create sufficient jobs to satisfy hundreds of thousands of school leavers each year."

In the face of the deteriorating economic situation at home and mounting external debt abroad, President Babangida had to act quickly and decisively. While he got the nation debating whether or not to accept a loan from the International Monetary Fund with its notorious conditions, the President worked quietly with the World Bank and other groups of creditors for a rescheduling of the country's massive debt of over $20 billion and thereby restore normal trading relations with Nigeria's overseas partners. The price was the programme of belt-tightening known as the Structural Adjustment Programme launched in the middle of 1986. The *Financial Times* described it as,

> "... one of the most radical of its kind in Africa. Not a sector of the economy has been left untouched: all but a handful of price controls have been lifted; fertiliser and petroleum subsidies cut; rice and wheat imports banned in an effort to encourage production of locally grown staple foods ... Up to 100 state-owned companies are to be privatised and Government is committed to ceilings on domestic borrowings.
>
> "At the heart of the programme was the Government's introduction of a two-tier exchange rate which not only led to an effective 66% devaluation of the naira but ended, at a stroke, the old import licensing system, one of the most serious areas of corruption and

> the most frequently used form of government patronage. Under the new system, available hard currency is auctioned each week and an import licence is automatically issued to the successful bidders."

The aims of the Structural Adjustment Programme (SAP), as they emerged, may be summarised under four broad headings:

> to restructure and to diversify the productive base of the economy in order to reduce dependence on the oil sector and on imports
>
> to achieve fiscal and balance of payments viability
>
> to promote growth while controlling inflation
>
> to promote efficiency and reduce unproductive investments in the public sector while intensifying the growth potential of the private sector

The main concern here is with the effect of SAP on industrial performance as reflected in the cement industry in general and WAPCO in particular in terms of capacity utilisation, capacity protection, expansion and new investment. So far, the impact has remained low. The intention to optimise capacity utilisation, expand plants or start new ventures has been frustrated by shortage of foreign exchange, inflation caused by the devaluation of the naira; high interest rate; tariffs on imported raw material; shortage of spare parts and other material such as cement paper bags, etc.

Diversification: Portland Electrical Repairs

One of the ways in which WAPCO has tried to adjust to the changing economic situation by diversifying out of its resource skills was by the establishment of the Portland Electrical division. In the context of the Structural Adjustment Programme this is perhaps an ideal example of the way in which established Nigerian companies can demonstrably assist with import substitution and the consequent saving of vital foreign exchange. In fact WAPCO was really ahead of its time in this regard as the beginnings of this activity were established in the mid seventies and were born out of a combination of necessity and initiative. The constant interruptions of NEPA power supply caused enormous damages to Ewekoro's electrical motors. Tony Jackson can remember writing to John Milne saying that the average expectancy of life of a WAPCO electrical motor was less than

Ron Mitten, the first general manager of Portland Electrical Repairs, with Col. Diya, the Governor of Ogun State, at the official opening of the new PER workshop at Shagamu

Another successful PER rewind being loaded

that of a British Army subaltern in the trenches in the First World War, i.e., three weeks. The rewind skills were not available within Nigeria and so WAPCO was faced with the costs of direct replacement or refurbishment overseas. Under the supervision of Ewekoro's then chief electrical engineer, Norman Butterworth, a rewind capability was established. Patrick Oyekanmi, one of the first employees he trained in rewind technology, was the current chief electrical engineer at Ewekoro. Over the next few years this capability was developed so that by the early eighties it was possible to recognise the potential of a separate business activity which would be capable of providing services to third parties. Accordingly a stand-alone facility was established at Shagamu.

Formally, Portland Electrical Repairs began life as a separate division in June 1982 under the general managership of an expatriate, Ron Mitten. Its aim was to expand the existing motor rewinding business, which was servicing the 2,000 electrical motors in WAPCO, to exploit the market for the rewinding, repair, refurbishing and servicing of electric motors and transformers. Initially the division leaned heavily on the advice, seconded help and the supply of special materials from the UK company, British Electrical Repairs Ltd (BERL), who had been operating in the UK and Europe for over 50 years.

The division was an immediate success and grew rapidly from a modest turnover of N187,000 and profits of N20,000 in the six months it operated in 1982 to sales of N643,000 with profits of N185,000 in 1983 and sales of N1.15 million and profits of N438,000 in 1984. By then it was serving major Nigerian companies, such as Lever Brothers Nigeria – its first customer – Dunlop Nigeria, Metal Box Nigeria, Machinery and Electrical Equipment, Nigerian Electrical Development Company and African Petroleum. The timing of the setting up of the division could not have been more apposite with the emphasis on the limitation of imports and the preservation of foreign exchange. Every motor rewound by PER was a motor that did not have to be replaced by an imported one or a motor that would have had to be expensively rewound abroad.

On 18 January 1985 a new workshop, constructed and equipped at a cost of N400,000, was opened by the Military Governor of Ogun State, Brigadier Oladipo Diya, and by this time the division was servicing 180 customers throughout the country. Later in 1985 another workshop was opened in Warri in Bendel State to serve local needs, primarily the oil industry. By this time the company's trucks were organised to bring customer's machines to the workshop at Shagamu from Ibadan and Lagos and the division was coping satisfactorily with alternators as large as 2 megawatts

By 1987 Ron Mitten had returned to the UK to set up his own business and was succeeded by Pat Roberts. Roberts proudly boasted in the annual report to shareholders of three types of specialist repairs – for light motors where a fast turnover was expected and

E A Aduwo – general manager, PER

given, for intermediate to heavy motors and alternators where turnaround could be expected in 14 to 28 days, and for bar wound motors and Megawatt alternators where replacement overseas could cost as much as N1 million. He also calculated that in its five years of operation the division had saved foreign exchange of more than $30 million.

In 1988 the expatriate Pat Roberts handed over as general manager to E A Aduwo and in that year sales reached N5 million and profits N1.56 million. By 1989 the 12 men who had been employed in 1982 had expanded to 159, turnover reached N11.3 million and pre-tax profits N3.2 million. The branch at Warri in Bendel State had been joined by another at Kaduna in the north. The workshop commissioned at Shagamu in 1985 had expanded to 15,000 square feet and the division's customer list had risen to 550. It included almost every major company in Nigeria – Guinness, Nigerian Breweries, Standard Breweries, Premier Breweries, Sona Breweries, Dunlop, T and E (a division of UAC), Nigerian Textile Mills, Nigerian National Petroleum Corporation, Shell, Santa Fe,

Relaunch of Sandtex Trowel and Cemwash – Dr St Matthew Daniel, the commissioner for commerce and industry, Lagos State

Paterson Zochonis, Coca Cola, A G Leventis, Pfizer, John Holt, the Federal Superphosphate Fertiliser Company and all the cement companies, water boards and glass industries. By now the division was a leader in the field of servicing high tension motors and alternators. Early in 1990 a workshop was opened specialising in the repair of air-conditioner compressors and refrigeration equipment and plans are well established to set up a transformer repair workshop.

With the country's continuing need to restrict imports and preserve foreign exchange and with NEPA raising their rates by up to 1,000%, thereby making everyone keen on generators again, the future for Portland Electrical Repairs looks very rosy indeed.

Portland Paints and Products

PER was not WAPCO's first diversification. At the end of 1973 WAPCO acquired Cement Paints Nigeria which was owned by APCM. This company had been originally set up in Eastern Nigeria to manufacture Permacem and Snowcem. Its operations were overtaken by the events of the Civil War. Its principal pieces of equipment, including the mill, were bravely and ingeniously rescued and brought to Ewekoro. This operation was masterminded by Jim McColgan, the current chief executive of Blue Circle Industries, and Tony Keeble. Cement Paints was absorbed by WAPCO to become the special products division initially under Tony Keeble and later Roy Hughes. It responded to the buoyant building conditions of the seventies by expanding its range of products and also embarked on marketing a range of complementary imported products.

WAPCO set up the division called Special Products to sell mainly imported building materials, such as Snowcrete, Snowcal and the Sandtex range from Blue Circle as well as scaffolding, electrical and plumbing items. In 1976 its turnover was N3.72 million on which it made a profit of N657,000. In 1977 its turnover increased to N4.85 million, but profits declined to N413,000 and in the following two years on sales of over N5 million it made losses of around N800,000. In 1980, therefore, management decided to turn it into a separate division called Portland Products. By this time import restrictions had made it necessary for the division to manufacture locally as many products as possible and it entered into a technical agreement with Blue Circle to manufacture under licence the Sandtex range: Sandtex Trowel, Sandtex Matt and Sandtex Textured. Under its general manager, J R Hughes, the new division made profits of N408,000 in 1980. During 1981 it began to manufacture paint and acquired the agency to market the sanitary ware of Armitage Shanks, a British company recently bought by Blue Circle. It did not have a good year in 1981 losing N78,000, mainly thanks to delays in obtaining imported raw materials, and losses at the docks through theft. By 1982, however, profits were N850,000 and would have been higher but for the difficulty in obtaining import licences for the Armitage Shanks range. New products had been developed specifically for the Nigerian market – a roller textured decorative coating, Bluetex, and a high quality emulsion paint, Vinyl Matt Emulsion. By this time Portland Products were used nationwide, recognised as having properties which preserved the appearance of buildings and also fought against fungal and algal growth. The factory was at the Ewekoro works and the administrative and sales offices at Ikeja near Lagos. But the problem of import licences continued, affecting raw material supplies and making it absolutely impossible to import and distribute plumbing fittings and flooring materials as well as Snowcal and Snowcrete. In the annual report to shareholders there was the usual plea to government,

> "The products of the Division are in great demand for Government buildings etc and we will be enabled to continue to make our valuable contribution to national development if we are given some import licences *from time to time* to meet our needs."

Distributors, PPP, from zones throughout the country at a meeting called by the company

By 1985 sales had been forced down to N4.3 million from its peak of N9.3 million in 1983. Again it was the lack of raw material import licences that was to blame. During the year the company – now a limited company, Portland Paints and Products (Nigeria) Ltd – had to cease production of the Sandtex range altogether.

By early 1977 the company had established six depots scattered throughout Nigeria, at Ibadan, Benin, Aba, Kaduna, Kano and Jos. In 1986 sales had slumped to N2.3 million as no import licences had been obtained for two years before 1986 and only ten per cent of the required N5 million of import licences in that year.

We have already seen the effect of the introduction of SFEM in September 1986 on the current operations and for PPP it also meant that raw materials could be imported again though at a high cost. In 1987 sales leaped to N7.1 million though profits were only N250,000. The improving trend continued in 1988 when T O Agboola took over as general manager from David Oni, who had been manager since 1984. The foreign exchange position was easier, allowing a more stable supply of raw materials. Sales doubled to N14 million and profits leaped up to N1.5 million. By this time many others in Nigeria were producing paints too – the company calculated there were 120 manufacturers (registered and unregistered) – and was fully aware of the competitive market in which it operated. Paint manufacture was "low tech" and the cost of entry into the market was low. Furthermore because of the

Akin Fadoju, company secretary and legal adviser (right) with Segun Agboola, general manager, PPP

straitened economic circumstances in Nigeria many people were taking the short-term view of looking at price rather than quality. PPP sold on quality and began to advertise aggressively using both television and radio. By 1989 the company had increased its depots to eleven adding Lagos, Abeokuta, Akure, Onitsha and Enugu to its original six. Sales jumped to N23.3 million and pre-tax profits to N4.1 million.

The company installed a unique manufacturing plant at the Ewekoro paints factory which produces a range of coloured glass aggregate from waste white glass. This aggregate is an essential material in the manufacture of Sandtex Trowel, which like many of the other products in the PPP (Nigeria) Ltd range contain approximately 70% of local material input. Sandtex Trowel, Sandtex Finebuild and Cemwash were re-launched successfully in 1989. PPP (Nigeria) Ltd is the leading producer of external decorative finishes in Nigeria and plans to complement the range by introducing gloss paint in 1990.

10

THE WAY AHEAD

The cement situation in 1989

In 1989 the familiar problem of demand far exceeding supply was still distorting the cement market. In February the *Daily Star* summed up the situation in an article under the title, "The gold called cement".

It traced the history of the cement industry back to the early seventies when the Gowon administration had reacted to the cement shortage by a massive import programme. That had led to near paralysis of the country's ports. By the mid eighties the state of Nigeria's economy ruled out direct imports of cement and shortage of foreign exchange was hampering indigenous production through the lack of spare parts which had to be imported. As a result the badly run cement factories were barely functioning and even a well-run plant such as Ewekoro could not run at full capacity. In such a situation the problem that had dogged the trade throughout its 30 year history inevitably reared its ugly head again. As the *Daily Star* put it,

> "In this type of situation where demand outstrips supply, middlemen usually swoop in for the kill. They are now having a field day, exploiting the situation to make huge profits for themselves. A situation where a bag of cement sells for about N45 in the open market, as against the factory prices of N19, calls for urgent actions."

That was early 1989 and by mid-year the situation had changed little except that prices had continued to rise with general inflation. The press were still calling for control of the distribution system so as to curb the excesses of the distributors. This is the *Daily Champion* in August 1989,

> "The cement companies, in collaboration with the government, should explore the possibilities of decongesting the product distribution outlets to ensure that cement gets to the ordinary consumers at marginal profit."

In a speech at a dinner party in September 1989 to celebrate the tenth anniversary of the Ashaka Cement Company, Akin Ogunleye, the chief executive of WAPCO and chairman of the Cement Manufacturers Association of Nigeria, highlighted the industry's and in effect the country's problems. He showed how the country's plants were producing way below their installed capacity,

Cement related facts facing the industry

Local capacity utilisation and imports (million tonnes)					
Year	Installed Capacity	Actual Production	% Utilisation	Imports	Total
1981	4.8	2.9	61	4.8	7.7
1982	4.9	3.2	66	4.2	7.4
1983	4.9	3.0	61	3.7	6.7
1984	5.4	3.0	56	1.3	4.3
1985	5.4	3.4	62	0.8	4.2
1986	5.4	3.5	65	0.8	4.3
1987	5.4	3.4	63	0.7	4.1
1988	5.3	3.4	64	0.8	4.2
*1989	5.2	3.0	58	0.7	3.7

*Based on half-year results

He showed that, compared with the rest of the world whose production was nearly 90% of installed capacity, Nigeria was performing badly and everyone was losing out as a result – government, industry, the investor and the consumer. In spite of this Ogunleye showed how the cement industry was benefitting the country in eight major ways,

1 Over 30 years the local production of cement had led to an enormous improvement in the standard of Nigeria's housing. It had also led to the growth and improvement in standard of public service buildings

2 Cement manufacture had made a significant contribution to the growth of Nigeria's Gross Domestic Product

3 Cement manufacturers employed 9,000 people directly and thousands more indirectly

4 Management of cement had meant the integration of other activities which saved foreign exchange such as the production of Kraft paper bags

5 The manufacturers made a significant contribution to government revenues through the payment of taxes, much more so than cement importers "because many of them are small, poorly organised businesses which easily escape the eye of the taxman especially as they are often not audited." Furthermore the government benefitted from dividends from its investments in cement manufacturing companies

6 Nigerian industry had benefitted from both the technical and manpower skills from Western Europe and by 1989 had been able to substitute Nigerians for expatriate managers so that no more than 2% of the cement workforces were now expatriate

7 There had been enormous savings in foreign exchange growing from N123 million in 1983 to N675 million in 1988

8 Cement manufacturers had been very responsible in their attitude to the local communities in which they operated. This was apparent in their approach to dust control, the awarding of scholarships to children of their staff, their response to appeals for charitable donations and their contribution to local welfare whether through the building of community halls and roads to remote areas or through the setting up of local sporting facilities.

Structural Adjustment Programme

Ogunleye acknowledged the merits of the Structural Adjustment Programme (SAP) in its stated objectives but he wondered whether in practice the government's actions were hindering the achievement of some of the objectives. The lack of foreign exchange for vital spares was preventing local manufacturers from improving the base of the economy and redressing the balance of payments deficit. The government's use of middlemen in purchasing was not helping "non-inflationary" growth or improving the efficiency of the public sector.

Specifically cement manufacturers not only could not expand their production but they could not even operate to their present capacity owing to the lack of foreign exchange necessary for the purchase of vital spare parts. The depreciation of the naira had been so large that development plans for the future – inevitably dependent on the purchase of plant and parts from overseas – had been abandoned or severely curtailed for the foreseeable future.

All the cement manufacturers were facing major delivery problems because of the unreliability, poor quality and sharply escalating prices of the only Kraft paper bag producer at Jebba. To add to this problem the supply of fuel oil had become erratic. This was scarcely comprehensible and certainly not excusable in a country whose major – some would say only – indigenous raw material was oil. Other cost increases faced by the cement manufacturers related to finance – with lending rates as high as 27% – electricity where the rate increase in 1989 was 1,000%, and raw material royalties to the government.

How was Shagamu doing after ten years of production?

Despatches had climbed steadily from 325,000 tonnes in 1978 to 610,000 in 1979, 628,000 in 1980, 707,000 in 1981, 878,000 in 1982, dropped back to 785,000 in 1983 before surging to 1,009,000 in 1984. They had dropped in 1985 to 947,000 and again in 1986 to

Despatches of cement, Ewekoro 1960-89

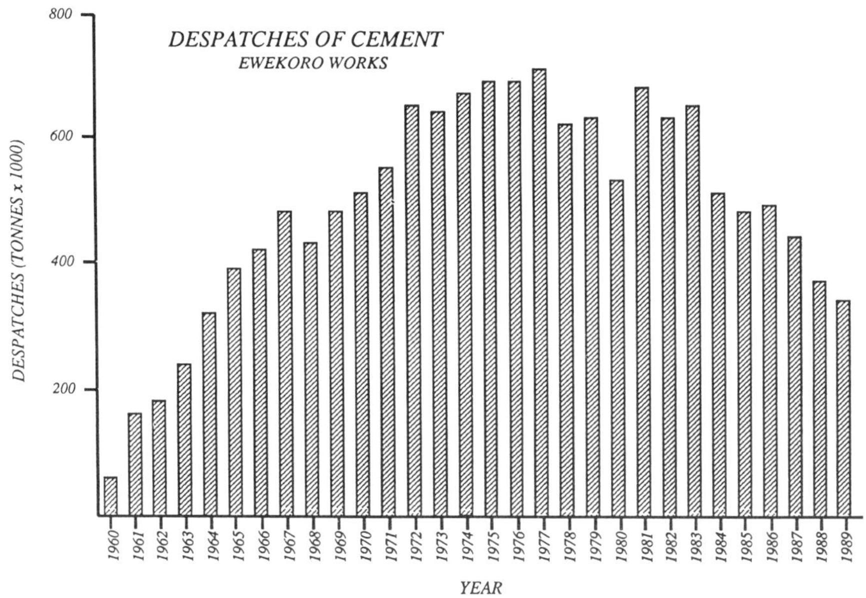

Despatches of cement, Shagamu 1978-89

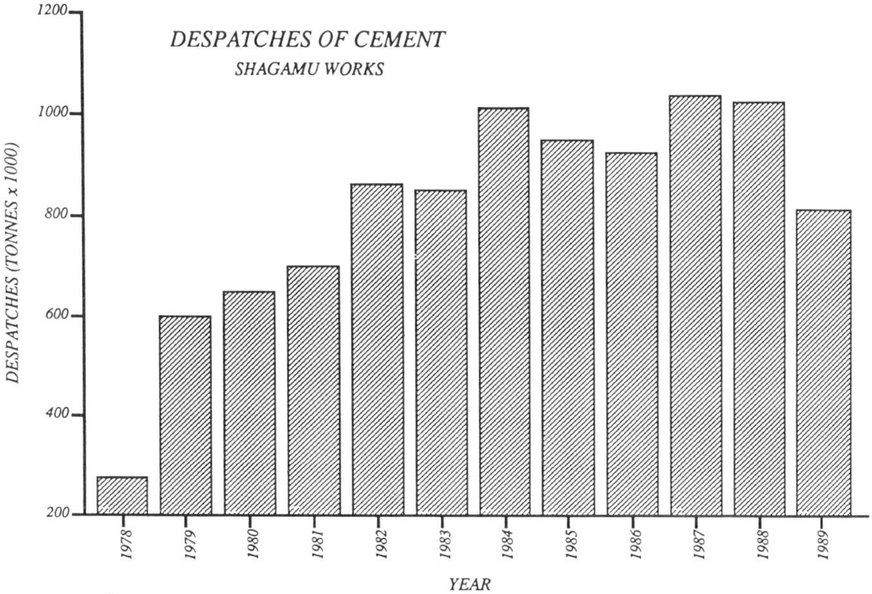

Despatches of cement, WAPCO 1960-89

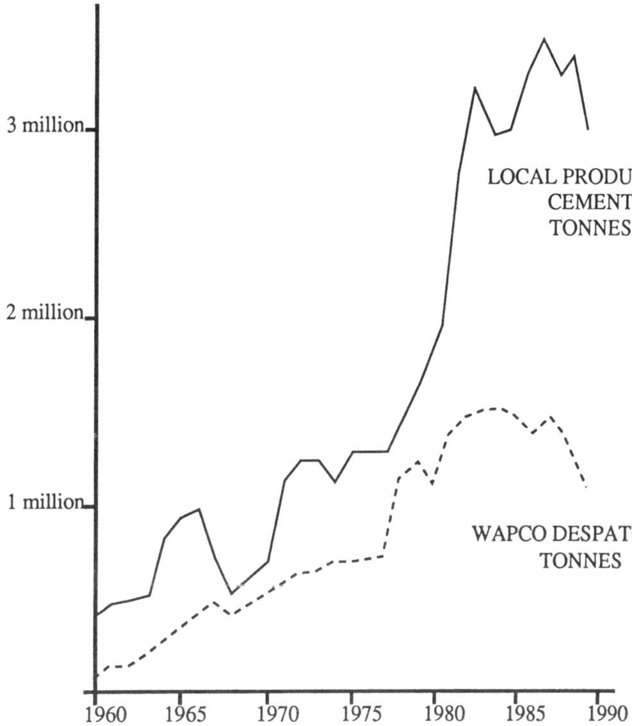

921,000 before surging again to an all-time record in 1987 of 1,030,000 tonnes, a level that was virtually maintained in 1988.

Johnson Otubaga was transferred from his position as general works manager at Ewekoro to take over control at Shagamu in 1987. He highlighted some of the aspects for the future of Shagamu,

> "Before the factory was sited at Shagamu a geological survey was carried out. This demonstrated that there were substantial deposits available and these were proved qualitatively and quantitavely to a scale that would justify the establishment of an operation of the originally designed capacity for an economically viable number of years. Blue Circle's geologists returned periodically to update the quarrying plan and to investigate further potential. Recently this latter need required confirmation in order to justify the construction of a third kiln. Adequate limestone deposits have now been proved to support a three kiln operation for 35 years"

The Shagamu factory has worked extremely well and has integrated with the local population with very few problems. One problem that had been a cause of complaint from the villagers nearby was the damage and noise level caused by the use of explosives in the quarry. Advances in quarry equipment technology have made it possible to reduce the level of explosives through the purchase of two RH75C excavators from O and K Ltd and an impact hammer thereby cutting down the use of explosives from 10 tonnes to one tonne a week. This brought several substantial savings in explosives, manpower who could be retrained and a reduced level of complaints. The new equipment was commissioned before an audience of local dignitaries who appreciate the efforts made by the company to be "good neighbours".

The effect of the sharp devaluation of the naira can be seen in the replacement cost of vital pieces of equipment. 35 tonne Terex dump trucks which cost N200,000 in the early eighties and N500,000 in 1988 cost N3.5 million by the end of 1989. A replacement radiator now costs almost as much as a complete truck in 1983.

A recent difficulty arose in respect of paper bags. WAPCO had for many years purchased its paper bags from local converters who manufactured bags from imported sack Kraft, one of which, Nigerian Kraft Bags Ltd, is now partially owned by WAPCO. Nigerian Paper Mills at Jebba, which produces Kraft partially from indigenous pulp, has a lower quality and is as yet an unreliable supplier. As a result of

The Alaperu of Iperu pouring a libation to commission the MH6 'woodpecker' impact hammer

An RH75C face shovel excavating and loading limestone in Shagamu quarry

The MH6 impact hammer

a shortage of paper sacks Shagamu despatches were curtailed by 200,000 tonnes in 1989 and the company was forced to spend many millions of naira to import sack Kraft paper from overseas suppliers.

The gas pipeline is fast approaching Shagamu and supplies should be available in 1992. Highly complex negotiations have been carried on by WAPCO with the Nigerian Liquid Natural Gas Company to ensure consistent and economic supplies for the foreseeable future and to minimise the costs of conversion from fuel oil. Complete plans for the installation of a third kiln with additional 3000 HP raw mill and cement grinding mills have been drawn up. However, under present restrictions on foreign exchange availability it is difficult to

Braimoh Musa – President of the union, National Union of Chemical and Non-Metallic Product Workers

predict when the company will be able to implement this plan to increase production by 450,000 tonnes per annum.

Human relations in WAPCO

In its 30 years of independence Nigeria has suffered, like many other countries both developed and emergent, a number of problems in industrial relations. WAPCO has not been completely immune but has an excellent record in its relationships with its staff. In his 1988 New Year message Chief Ogunleye congratulated all the company's personnel for their restraint in a year which had seen considerable strife throughout Nigerian industry,

> "An area in which we have congratulated ourselves over the year has been the continued industrial peace within the company. Because of the economic conditions obtaining in the country we had a period of relative

Chief David Akinfolarin (left), head of personnel, Patrick Gorman, works manager of Ewekoro, and O O Ipeaiyeda, financial controller

concern over our ability to maintain the peaceful atmosphere this year. The problem was mainly one of inadequate communication and understanding. Our staff associations used the established system of pursuing their genuine demands and in the end matters that could have created industrial strife were amicably resolved."

WAPCO has co-operated closely with staff associations and unions throughout its history. Braimoh Musa, President of the National Union of Chemical and Non-Metallic Products Workers, began working for WAPCO when he left school in the mid-sixties. He discovered he had a flair for trade union activities and became a shop steward in 1965, rising in the union hierarchy until he became President in 1978. Over the period he has established good relations with the management and has helped to improve working conditions for his members. For example, he helped establish a safety committee in 1970. He has also negotiated improvements in his members' fringe benefits – the WAPCO retirement benefit scheme is the best among the 250 factories in which his union is active.

As far as Musa is concerned the WAPCO management has been, and has been seen to be, fair and firm. This has helped him to control his unions and there has been no strike in the company since the early seventies. He admires the way the company responds to the changes in society and the way it gives preference in employing

G A Akinlolu, president of the SSSA addressing the AGM of the Association

locals at Ewekoro and Shagamu unless it needs to import particular skills. The company also appears to him to employ, if possible, the children of retired employees. As a result of its approach the company exerts a high level of discipline and commands great loyalty – Musa can remember going in to work in the factory himself on Christmas Day, 1978. His comrade, Shakim Rasaki, chairman of the same union as Musa for the company's three locations – the headquarters, Ewekoro and Shagamu – agrees with him. He feels the union and management have progressed together in their relations. The management have an open door policy and he sees the management and the union as "partners in progress".

G A Akinlolu, the branch chairman of the Supervisory and Secretarial Staff Association of WAPCO, looks after 250 staff in the company. He feels that WAPCO is a leader compared with other companies in its treatment and relationships with its staff. He meets regularly with David Akinfolarin, the personnel manager, and if negotiations become deadlocked, refers to the managing director. He has only had to do that twice since 1987. At the company's head office, Alhaji Adewumi, the head office accountant, is president of the Management Staff Association, and discusses the problems and conditions of services of the 200 managerial staff in the company's three locations. He also negotiates with Akinfolarin and with the chief executive Akin Ogunleye. Ogunleye is himself a former president of the MSA.

WAPCO therefore has many lines of communication to its staff and since the seventies, according to Layiwola Obanla, the assistant

Alhaji Gani Adewunmi, immediate past MSA president

secretary of the MSA based at Ewekoro, has pursued a more open style. Union agitation for Nigerianisation within the company preceded the Indigenisation Decree and led to the Nigerianisation of some key positions in the factory. For example, David Adeyemo was promoted to works manager. It has also improved its relationship with the staff by carrying out a regular monitoring of their progress. He remembers that Tony Jackson was instrumental in introducing job evaluation into the company and can quote from a letter written to him in 1975 by Jackson,

> "Dear Mr Obanla,
>
> "Whilst I was on leave I heard from Mr Sams [general works manager] of the successful introduction of job evaluation into the company. I would like to

David Adeyemo, recruited as early as 1957, rose to become commercial director of WAPCO before his tragic death in January 1985

> congratulate you on the part which you and your members together with your union secretary played in its successful introduction."

In 1989 WAPCO introduced a profit sharing scheme. Musa was fulsome in his praise.

"It is not often that workers praise Management, but when there are compelling reasons to do so the Union should not fight shy but to show appreciation and gratitude to a magnanimous management.

"There is no doubt that all employees in WAPCO can joyfully count their blessings since 1987 till now. At the end of last year, WAPCO Management became the 1st NUCANMP Company to introduce 'Profit Sharing Scheme' – a Scheme where every confirmed employee is entitled to a percentage in cash of the gross profit made annually by the Company. This is irrespective of whether or not the employee bought shares in the Company.

"In May this year the 15% general wage increase and the vastly improved annual incremental rate were implemented to the letter. The Company's Salary Structure for the Junior Staff was elongated to fifty and eighteen scales respectively, to ensure that reaching the end of one's scale early is eliminated. This, no doubt, was a rare concession made by management. We are pleased therefore that the great sacrifice made by the entire workers for the growth and survival of the Company is appreciated. In the concessions galore, the Company has also augmented the supplementary medical subsidy by additional N120,000 this year to reach a total of N350,000.

"It is however imperative for me to mention that the rapport between the Union Leadership and Management was largely responsible for the Company's progress and performance. I wish that the high standard of industrial relations practice in WAPCO would continue to be demonstrated by the two social partners."

WAPCO has also worked hard to build loyalty in its staff by introducing a long service awards scheme. First introduced in 1970 with a ten years service award, it was modified in 1979, 1983 and 1989 to introduce 20, 25 and 30 years service awards. In February 1989 the scheme was reviewed and the following awards agreed,

10 years – N250 plus Certificate

15 years – N350 plus Certificate

20 years – N500 plus Certificate plus a small office size refrigerator

25 years – N750 or one month basic salary (whichever is higher) plus a Certificate plus a gas cooker or a gold wristwatch

30 years – N1,000 or one month basic salary (whichever is higher) plus a Certificate plus a 20″ television

The company scholarship scheme, also first introduced in 1970, now provides 20 scholarships a year, all for secondary education, 16 of them reserved for children of junior staff. In addition two are provided at university level for the study of engineering. Another benefit offered by the company to its employees is comprehensive medical facilities whereby both works have clinics with nursing sisters and doctors are retained at Shagamu hospital. At the existing head office there is no room for a clinic and the company therefore retains two doctors at hospitals in Lagos. Other benefits include a gradual improvement in leave allowance, an increased number of salary steps within grades and an improved pension scheme. Finally, since the tragic murder of David Adeyemo, the company's commercial director, in January 1985 as he arrived at Elephant House, the company has tightened its security arrangements to give full protection to its staff.

Yemi Ajayi – public relations manager

Public relations at WAPCO

Part of WAPCO's concern for its employees, consumers and suppliers is manifest in the effort the company's public relations and advertising department exerts to communicate with all its internal and external contacts. It has also, with the aid of the MSA, organised seminars to bring together cement manufacturers, the public and government officials to discuss issues of public interest. Internally *Elephant Newsletter* is distributed monthly to all managerial and supervisory staff and *Elephant News* is distributed quarterly to all members of the staff. Besides these publications the public notice boards throughout the company's locations are used to disseminate information and the PR department talks to staff associations and the trade unions on a group basis. Externally the department keeps in constant touch with the media and in one arm or another the company is mentioned almost every day. The company has always seen sport as an important area of communication and the PR department is actively engaged in promoting the company's own Sports Day and golf tournament as well as sponsoring outside squash and table tennis tournaments.

Akin Ogunleye with Bryn Jones, the company engineering manager, and James Obaseki, the district director of NEPA, on the 9th at the Blue Elephant golf club, Ewekoro

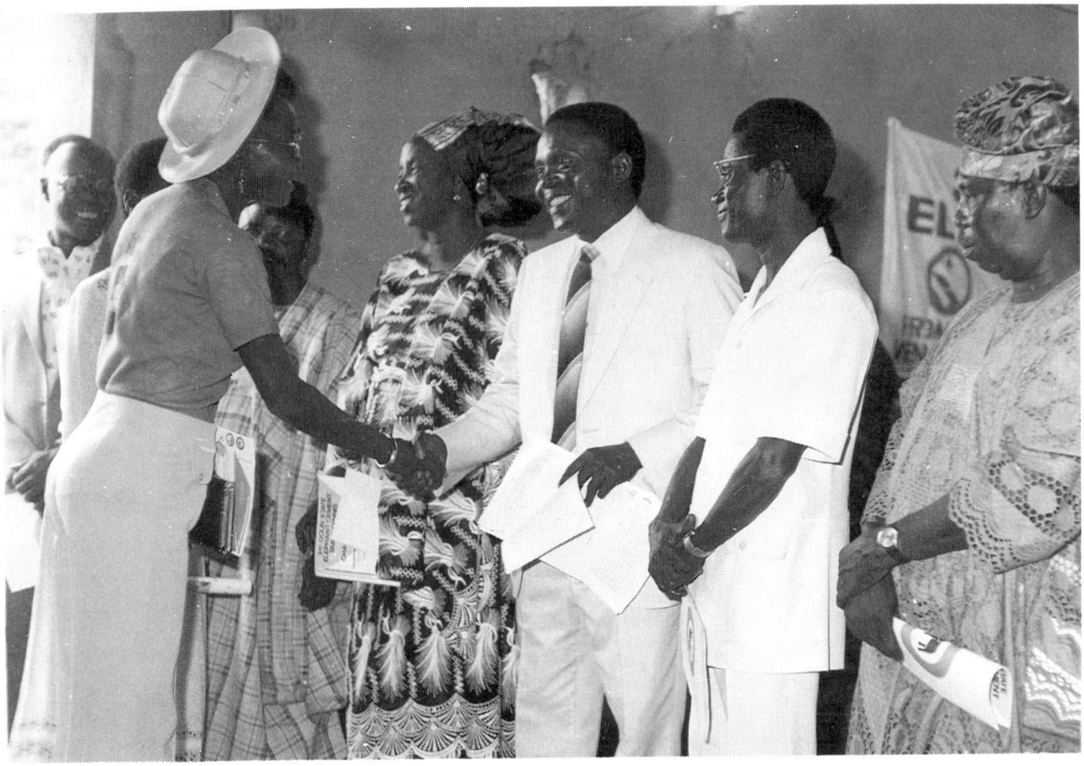

At a WAPCO sponsored table tennis tournament, the Hon Commissioner for Information, Sports and Social Welfare, Mrs Morenike Onanuga, shakes hands with Olu Adelaja, chairman of the Ogun State table tennis association

A winner at the tournament

The
WAPCO sports day

Training at WAPCO

We have seen that training was given a high priority by the company throughout the sixties and this approach has been intensified in subsequent years. In June 1968 a Training Centre, soon to become residential, was established whose primary aim was to retrain new recruited artisans and craftsmen from technical colleges whose previous training experiences were not completely relevant to factory requirements. With the opening of Shagamu in the mid seventies the training centre activities were expanded. Apprentices were admitted from Form 3 and 4 secondary school leavers and put through a four year programme. One of the problems was that no sooner had WAPCO trained these apprentices than other employers poached them. WAPCO responded by increasing the intake and establishing a staff training and development department. Within this department, management courses, both within and without the company, were organised for all categories of staff. In the case of senior staff several have been sent overseas for advanced courses in management and cement technology.

In 1978 the company conducted an experiment by taking on as production trainees six candidates with credits in chemistry, mathematics and English. This experiment was successful and some are now holding responsible positions in the company. 1989 began with the company employing 79 apprentices and during the year over 600 staff attended courses of one sort or another and 25 students attended courses ranging from two to six months.

The new Elephant House

Coinciding with the celebration of 30 years of production of cement in Nigeria, WAPCO is opening a new office complex in the Alausa district of Ikeja to replace its existing head office on Ikorodu Road in the Ilupeju area of Lagos although a sales office may be maintained in the old offices. The foundation stone was laid in December 1989 by the Governor of Lagos State, Colonel Raji Rasaki. The main contractor has been G Cappa Nigeria Ltd and the architectural consultants, Fola Alade Associates. Fola Alade, having done his basic training in Nigeria, completed his studies with a post graduate course at the Architectural Association school in Bedford Square, London. During the sixties and seventies Alade worked on various architectural projects for the state and federal governments including the re-designing and restructure of Dodan Barracks, the seat of the Government since the time of Yakubu Gowon. Between 1974 and

Chief Fola Alade, consulting architect for the new Elephant House

1976 he designed several overseas chancelleries including in Africa: Accra, Abidjan, Yaounde, Gaborone, Lusaka, Monrovia and Rabat; in Europe: the Nigerian Embassy in Bonn; in the Americas: the embassies in Brasilia and Mexico City and in Asia: in Islamabad, Pakistan.

In 1979 Alade retired from Government service and the following year started in private practice. Because he was still associated with the military governments, he found some difficulty winning contracts in the early eighties, but nevertheless was involved in projects with WAPCO, the university of Ibadan and the teaching hospital at the university of Ife. With the return of the military government he was asked to work on the ministry's offices at Abuja. According to Alade, President Babangida felt that good servants of the state should be rewarded.

In 1988 he was approached by WAPCO to work on their new office block. The overall brief was that the building should show the character of concrete and project the correct image of the company.

Jos Federal Secretariat – designed by architect Chief Fola Alade

The WAPCO board had been impressed by Blue Circle's new offices at Aldermaston in England and invited Alade to go and see them. He took the opportunity to talk to the architect, Richard Gilbert Scott. Alade was instructed to design a building which would show, as Aldermaston did, a sympathetic use of concrete. Clearly he could not imitate Aldermaston exactly – there was no lake, no natural woods, no vegetation, and more security was necessary. The original plan was for ten floors – this had to be scaled down to eight because of the proximity of the international airport. WAPCO will take two and control the conference centre while the other six are let out.

The way ahead

During 1989, among the developments at WAPCO with implications for the future, two may be mentioned. First A A Abidogun became chairman in June 1988, having been nominated to the Board as a representative of O'dua Investments Limited. He brought with him

Joe Makoju, recently appointed assistant managing director and chief executive designate of WAPCO

David Oni, general manager sales, and his wife, Deola

A A Ehindero, general manager – commercial

from Lever Brothers and Lipton Nigeria wide experience of other industries and great diplomatic talents. Although he modestly describes himself as a farmer, he has established a remarkable empathy with the Company and with its shareholders as was fully demonstrated in the very able and lively manner in which he conducted the Annual General Meeting of the Company. Secondly, Joe Makoju was appointed assistant managing director and chief executive designate with effect from January 1990. He was recruited

J O Uwasomba, non-executive director and representative of the Federal Ministry of Industries

in 1975 in the UK and spent two years training with Blue Circle Industries. After various technical and management posts, both at Shagamu and Ewekoro, he was sent again to the UK in 1988 to attend a senior course at Henley Management College as one of the candidates being groomed for succession to the post of chief executive. In 1989, he attended a nine month course at the National Institute for Policy and Strategic Studies at Kuru, near Jos.

The Nigerian business magazine *This Week* produced tables for

Professor A Soyode, non-executive director and representative of Odu'a Investments Limited

1988 for the top hundred companies in Nigeria. Excluding banks and oil companies WAPCO was ranked tenth in terms of turnover and sixth in terms of profits.

In 1960, WAPCO came into being as part of the conscious central planning of economic development by the government of the then Western Region of Nigeria. As WAPCO celebrates its thirtieth anniversary, the theory of economic development has gone full circle and government both at federal and state levels is pursuing an active

Gbenga Atoki, non-executive director and representative of Odu'a Investments Limited

policy of structural adjustment and privatisation, emphasising the role of the private sector and market forces. WAPCO has been fortunate from the start in its blend of local public capital with foreign private investment and technical expertise. The flexibility and liberal spirit with which this partnership has been pursued is such that WAPCO has been able to grow and to adjust to changing circumstances without undue strain or any evidence of structural imbalance.

Rodney Little – a non-executive director. After serving in Nigeria with the United Africa Company from 1957 until 1967 he joined Blue Circle in 1980 and has provided regular liaison between London and Lagos seeking solutions to commercial and financial problems

Initially the Western Region through the Production Development Board provided 39% capital and necessary political base. APCM, holding 51% was granted technical and management control to protect its majority shares. It appointed three of the six Directors, one of them being both chairman and managing director, with a casting vote. UAC holding 10%, along with APCM provided experience of the local market and international trade. Foreign capital restrained excessive nationalism, political infighting and the

Dr Paul Warburton, technical director, and his wife, Hazel

pressures for political patronage. Local participation ensured that even before the Indigenisation Decrees, WAPCO was a Nigerian company. WAPCO survived the crisis in the Western Region in 1962-66. In the process, and as befits the leading manufacturer of a scarce strategic commodity, WAPCO became more of a national Nigerian Company without losing its close relationship with its local communities and, through Odu'a Investment Company, the governments that have succeeded the former Western Region. In 1978, WAPCO became a Public quoted Company, with (WNDC)/Odu'a holding 26.8% shares; Federal Government 16.6%; Nigerian public 16.6%; (APCM)/Blue Circle Industries 33.4% and UAC 6%.

The success of WAPCO has called for more than the production of good quality cement. There is also the marketing and distribution, accord with local communities and public relations, training of staff and labour relations, investment and financial management, management of stores, foreign trade and foreign exchange in the importation of spare parts and foreign technical skills. WAPCO began

Mrs Tina Apatira – recruited by Tony Jackson she has performed a vital role as personal assistant to many of the senior executive directors

with an all expatriate management and technical staff and Nigerian labour. Trained Nigerian staff now manage operations of the Company in all divisions: production, maintenance, marketing, personnel, finance. The majority of the Board of Directors, including the chairman and the chief executive are Nigerians, while the non-executive vice-chairman, the deputy managing director and the technical director are expatriates. Its now acclaimed package of staff welfare, working conditions and training schemes were worked out in response to WAPCO's own needs and internal pressures, often ahead of public policies of indigenisation.

While there is a great sense of pride – fully justified – in WAPCO's achievements in its 30 years of cement manufacture in Nigeria, there is also a sense of frustration in that the company knows it could do so much better if only it could invest further to increase its output and meet more of the latent demand. We have read constantly in these pages of shortage of foreign currency leading to delays in importing vital spare parts and machinery. WAPCO is now working

Cementing a partnership

on a major piece of creative financial engineering whereby massive investment in its plants can take place. If successful this will lead to a large increase in cement output and WAPCO will be able to continue to contribute significantly to the development of the Nigerian economy as it has for the last 30 years.

APPENDIX I

WAPCO Chairmen, Vice Chairmen and Managing Directors

	Chairmen	Vice Chairmen	Managing Directors
1960–5	T R Chesterfield		T R Chesterfield
1966–71	A E Taylor		A E Taylor
1971-73	R Cheney		R Cheney
1973–6	A J Jackson		A J Jackson
1976–9	Chief O I Akinkugbe OFR		Chief A J Jackson OBE
1979	Chief O I Akinkugbe OFR		R Burnett
1980–1	S O Ige	C S O Akande OFR	R Burnett
1982–3	S O Ige	C S O Akande OFR	C Roots
1984–5	Otunba V A Oduntan	Chief A J Jackson OBE	C S O Akande OFR
1986	Otunba V A Oduntan	Chief A J Jackson OBE	Chief E A Ogunleye
1987–8	J F Ale	Chief A J Jackson OBE	Chief E A Ogunleye
1988	Prof M A Adeyemo	Chief A J Jackson OBE	Chief E A Ogunleye
1988–90	A A Abidogun	Chief A J Jackson OBE	Chief E A Ogunleye

APPENDIX II

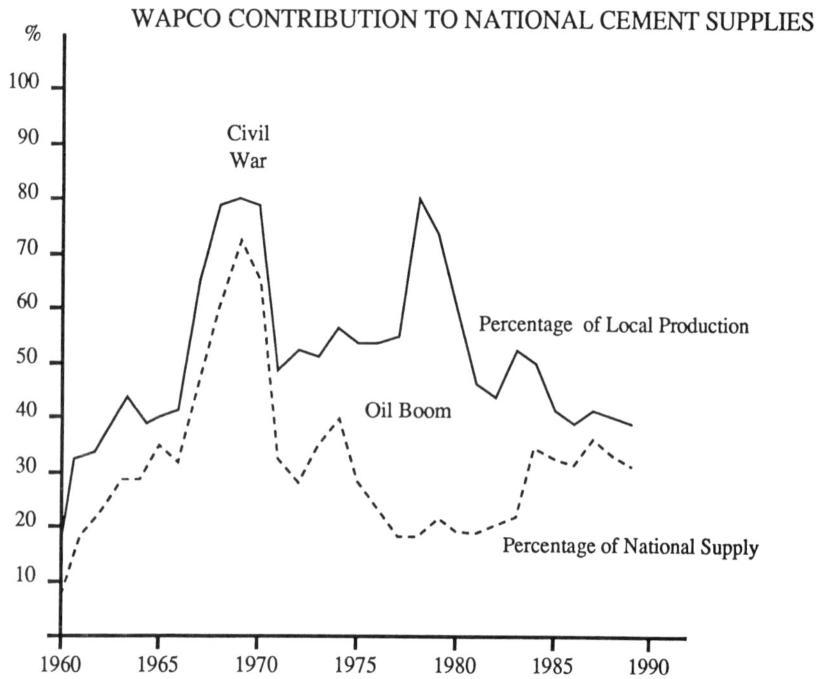

WAPCO CONTRIBUTION TO NATIONAL CEMENT SUPPLIES
%
100
90
80
70
60
50
40
30
20
10
Civil
War
Oil Boom
Percentage of Local Production
Percentage of National Supply
1960
1965
1970
1975
1980
1985
1990

BIBLIOGRAPHY

BIERSTEKER, THOMAS J *Multinationals, the State and Control of the Nigerian Economy* (Princeton University Press, 1987)

EKUNDARE, R O *Economic History of Nigeria* (Methuen, London, 1973)

KILBY, PETER *Industrialization in an Open Economy: Nigeria 1945-1966* (Cambridge University Press, 1969)

O'CONNELL, J C "The Political Class and Economic Growth" in *Nigerian Journal of Economic and Social Studies* March 1966, pp 129-40

PERHAM, MARGERY (ED.) *Economies of a Tropical Dependency: The Native Economies of Nigeria,* Vol ii *Mining, Commerce and Finance in Nigeria* (Faber & Faber, London 1948)

PUGH, PETER *A History of Blue Circle Industries* (Cambridge Business Publishing, Cambridge 1988)

TERIBA AND KAYODE (EDS.) *Industrial Development in Nigeria: Patterns, Problems and Prospects* (Ibadan University Press, 1977)

UGOH, S U "The Nigerian Cement Company" and "The Nigerian Cement Industry" in *The Nigerian Journal of Economic and Social Studies* March 1984 pp 72-91; March 1966 pp 97-112

INDEX

A page reference in italics indicates where an entry is illustrated